AF678288

PROCESS TECHNOLOGY TROUBLESHOOTING

Thomas D. Felder

Victoria College

E. Lamar Garrett

Editor, Process Technology Series

Prentice Hall

Upper Saddle River, New Jersey
Columbus, Ohio

To Bobbie Lee

Library of Congress Cataloging-in-Publication Data

Felder, Thomas D.
Process technology troubleshooting / Thomas D. Felder, E. Lamar Garrett.
p. cm.
Includes index.
ISBN 0-13-027937-4 (alk. paper)
1. Chemical process control. I. Garrett, E. Lamar. II. Title.
TP155.75.F47 2004
660′.2815—dc21

2002044552

Editor In Chief: Stephen Helba
Assistant Vice President and Publisher: Charles E. Stewart, Jr.
Production Editor: Alexandrina Benedicto Wolf
Production Coordination: UG / GGS Information Services, Inc.
Design Coordinator: Diane Ernsberger
Cover Designer: Rod Harris
Production Manager: Matthew Ottenweller
Marketing Manager: Ben Leonard

This book was set in Times New Roman by UG / GGS Information Services, Inc. It was printed and bound by R. R. Donnelly & Sons Company. The cover was printed by Phoenix Color Corp.

Pearson Education Ltd.
Pearson Education Australia Pty. Limited
Pearson Education Singapore Pte. Ltd.
Pearson Education North Asia Ltd.
Pearson Education Canada, Ltd.
Pearson Educación de Mexico, S.A. de C.V.
Pearson Education—Japan
Pearson Education Malaysia Pte. Ltd.
Pearson Education, *Upper Saddle River, New Jersey*

Prentice Hall

10 9 8 7 6 5 4 3 2 1
ISBN 0-13-027937-4

PREFACE

The Gulf Coast Process Technology Alliance leads the movement within the chemical and related industries to employ process operators who complete an education in process technology. The Alliance, with input from academia and industry, has specified the subject matter for use in the study of process technology. The specifications are very inclusive, permitting a broad adaption of course materials that complement the technology in use by potential employers. Petrochemical and oil-refinery operations are major targets.

The study of process technology troubleshooting in an academic setting is supported by two pragmatic concepts. First, the physical and chemical principles at work in the unit operations that make up a petrochemical plant are well defined. Industrial products are produced by the creative application of well-known process technology. Secondly, the process operator is trained on the job to carry out specific physical and chemical changes on selected feedstocks through the operation of a carefully selected mix of equipment.

The subject material is selected so that the novice process operator can put this knowledge to use as soon as he starts on the job. Emphasis is placed on how the process-operator job fits into the overall team effort needed to run a petrochemical plant safely and productively. An important responsibility in process troubleshooting in teams is effectively communicating with those assisting in the management of process problems.

Community colleges, with industry support, have introduced two-year programs that earn an Associate of Applied Science degree in process technology. Process technology troubleshooting well deserves a place as an advanced offering in the program. The integration of the process operator position into overall technical resources is a troubleshooting plus.

The student enrolled in process technology troubleshooting has completed a carefully chosen mix of academic and technical courses but likely has limited experience with plant operations. This creates a challenge: writing course material that simulates workplace conditions that the student can be expected to troubleshoot. The objective of this material is to offer simple process incidents with a sense of urgency for the students to solve and troubleshoot. A heated debate on probable cause within a team of student troubleshooters is a productive environment in which to teach the steps in the problem-solving process. Another reasonable goal is for course material to present the problem-solving process through exercises that the student can attack based on his previous studies.

The material selected here places emphasis on workplace organization, communications, teamwork, a workable problem-solving process, and problem prevention. Specific to the needs of a novice process operator as a troubleshooter in training, the material teaches the following:

- The responsibilities of the process operator as a troubleshooter on a technically oriented troubleshooting team.
- The application of process technology to the detection, definition, and communication of the effects of process problems.
- Typical problem categories called climates for problem development; an unexpected problem, to the dedicated troubleshooter, is simply an expected problem left undetected.
- The importance of the role of primary communicator of process conditions to other troubleshooters.
- A rational problem-solving process and how it is a natural part of a successful troubleshooting effort.
- Problem prevention, which is the ultimate accomplishment of the process operator as troubleshooter.

The masculine pronoun *he* and adjective *his* are used throughout the text to represent both sexes. The repetition of *he* or *she* in the sentence, "He or she make a good troubleshooter" is tiresome, so the masculine *he* has been selected to aid in readability.

Thomas D. Felder

CONTENTS

CHAPTER

1

The Troubleshooting Team

LEARNING OBJECTIVES

- See the importance of the process operator as a troubleshooter, since he is in the ideal position to monitor and report process conditions.
- Understand that the petrochemical plant is organized with the technical expertise to troubleshoot and solve production, mechanical, and technical problems. That expertise includes the position of the process operator.
- Recognize that the process operator as troubleshooter knows who the experts are, what they can do, and how to use their assistance in a troubleshooting team effort.

KEY TERMS

Teamwork
Troubleshoot
Productivity
Team, production
Team, mechanical
Team, technical
Troubleshooting team
Troubleshooter
Plant staff
Team, operations
Hazardous materials
Contract services
Contract worker
Work practice
Process upset
Shift
Discovery
Effect
Standard operating condition
Prime movers
Network
Yield
Utility
Certification

Teamwork
A shared effort by knowledgeable and skillful people to reach a common goal.

Troubleshoot
An organized effort to prevent process upsets through discovery of abnormal effects, analysis of abnormal effects for cause, and taking of action to eliminate cause.

Productivity
A comparison of performance to expectations, including utility, rate, yield, and labor.

Team, Production
Responsible for the operation of plant facilities.

Team, Mechanical
Responsible for the maintenance of plant facilities, including equipment, instruments, and electrical.

Team, Technical
The expertise needed to support the technology needs of the production and mechanical teams.

Troubleshooting team
A group of people selected for their knowledge and skills who observe, document, analyze, and correct abnormal process conditions.

Troubleshooter
One who seeks out unwanted process change and the causes behind it.

Plant Staff
Overall management of plant resources to meet business objectives.

Team, operations
The unified resources of the mechanical, production, and technical teams.

An industrial plant continuously executes interrelated activities that involve skilled personnel and the facilities they operate. All continuous operations depend upon *teamwork* using mechanical and technical resources to *troubleshoot* to detect and correct process problems that would otherwise penalize *productivity*. The process operator works closely with *production*, *mechanical*, and *technical teams* to detect and solve problems that would be impossible for him to solve alone.

The responsibilities of the process operator include the communication of process needs to assigned production, mechanical, and technical teams. The information must be documented in precise, technical detail, for there is limited opportunity for person-to-person contact. This is not a simple communications problem. Success is next to impossible if the process operator is not thoroughly familiar with the technical expertise available in the organization to assist him in meeting his responsibilities.

1.1 THE TROUBLESHOOTING ORGANIZATION

Industrial plants are organized into a matrix of *troubleshooting teams* with assigned members from the manager to the process operator (Figure 1–1). For example, a troubleshooting team may be composed of a supervisor and several technical specialists in a discussion with mechanical and production teams concerning a major process upset or a master plan for the annual turnaround. In daily plant life, the scenario is more likely to be a process operator, a mechanic, and a process control engineer mulling over what action to take on a worrisome process condition. If the three *troubleshooters* come up with a good decision on the corrective action they choose to take, a potential disruption in operations is prevented.

1.2 PLANT STAFF

The major troubleshooting responsibility of *plant staff* is to ensure that the *operations team* has a complement of able production, mechanical, and technical personnel qualified to solve and prevent process problems. Plant staff is not involved in the direct production of products but is responsible for overall plant performance.

The allocation of resources and the expenditure of capital are centered in plant staff. Plant staff is responsible for safety standards, public rela-

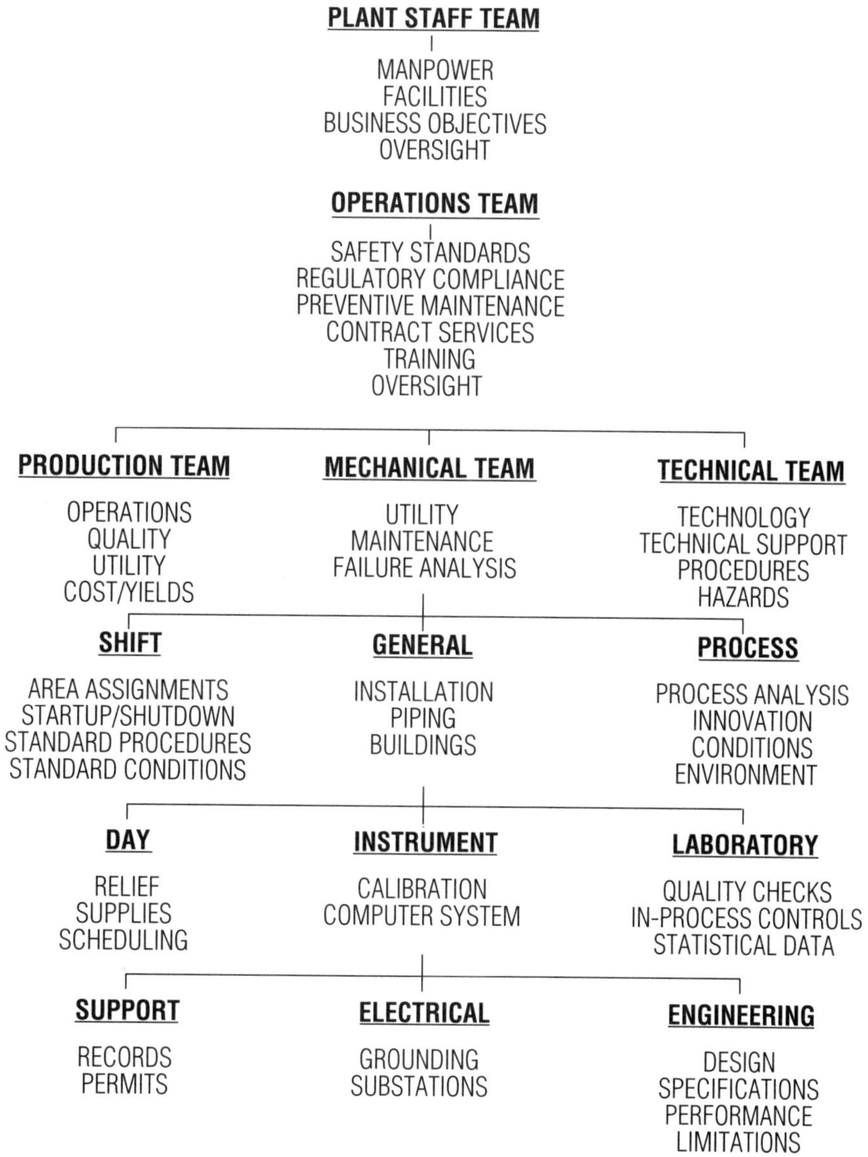

Figure 1–1
Troubleshooting Organization

tions, employee benefits, compliance with regulations, and plant security. In summary, plant staff provides the work environment, facilities, and human resources that are required to meet business objectives. All other plant activity is referred to as *operations* and is the responsibility of an operations team.

1.3 OPERATIONS TEAM

The operations team is a carefully crafted mix of production, mechanical, and technical personnel. The entire operations team is organized to support the production team in the continuous improvement of plant performance. The process operator is a production team member. He is expected to use all of the expertise contained in the operations team as the technical support necessary to do his job.

The operations team is presented here from the position of the field process operator, who is most likely to begin his career with on-the-job training in the field. The central control room and day operators are trained and have field experience.

1.4 PRODUCTION TEAM

The production team is charged with the safe operation of all plant facilities. The core responsibility is to safely and consistently meet production goals on rate, cost, and quality. The job often requires responding to special customer needs. Another important responsibility of the production team is the receipt, containment, and shipment of *hazardous materials*.

The production team is also the guardian of high standards in safety and work practices in all the process areas by all workers. The production team is held accountable for the shutdown and preparation of equipment for repair that minimizes risk to mechanics. *Contracting services* to an outside firm is common practice today. In such cases, the production team has the responsibility to explain the safety and work standards to all *contract workers*, since most are unfamiliar with the process.

Hazardous Materials
Process materials with toxic, flammable, or other properties that are harmful to human health.

Contract Services
Plant activities completed by outside workers under contract. The maintenance team is usually the coordinator.

Contract Worker
Outside personnel used to complete selected work under a contract.

Work Practice
An accepted way or method to complete a defined task.

Process Upset
A process unit operating unexpectedly outside standard operating conditions.

1.5 PROCESS OPERATOR

The process operator maintains standard operating conditions (SOC) through the use of standard operating procedures (SOP). This is a major responsibility, but he has other important work to do:

- He monitors all activity in his area, which includes assisting contract workers on safety and *work practices*.
- He prepares equipment for maintenance and follows the work of maintenance personnel.
- He anticipates process problems, using routine patrol to discover abnormal conditions in time to prevent *process upsets*.

- He communicates effects on process problems to others in the operations team with a need to know, which initiates technical assistance.
- He communicates the status of all activity on his *shift* for the benefit of his relief, which is essential to achieve a smooth transition from shift to shift.

Shift
A rotated work period needed to cover continuous operations.

In all aspects of his work, the process operator is a motivated troubleshooter. Upon *discovery* of an abnormal condition, the process operator first considers the safety of himself and other personnel at work in his area. He is alert to any deviation from a standard condition that could result in a process upset. He is expected to take corrective action on an abnormal condition through use of accepted procedures. His powers of observation, his process knowledge, and his omnipresence in the plant are key resources in the early discovery and prevention of process upsets.

Discovery
The first step in the problem-solving process. Collect the effects of a problem into a logical order.

A key responsibility of the process operator as troubleshooter is to collect and document precise and accurate data on a process problem. The *effects* that he documents on a problem are crucial in identifying root cause, which is an essential step in the problem-solving process if reoccurrence is to be eliminated. He writes down preliminary answers to the questions *where*, *when*, *what*, and *how*. He describes the action taken on a problem and who participated in that action. Everyone involved is a valuable source of information to a troubleshooting team.

Effect
A measured, observed, or sensed abnormality in the interaction of people, places, and/or things.

A pocket notebook and pencil are essential equipment for any serious troubleshooter and are essential to accurate recall of effects later. Odors, sounds, spills, heat, and human error are important effects of a problem that no instrument can measure and record. Time, place, and personnel involved are also crucial data in the reconstruction of all aspects of an abnormal condition and the corrective action taken at the time. Accurate recall of numerical data and the need to place events in a precise time frame are supported by good notes.

The process operator is the most important member of the operations team as a troubleshooter in a number of areas:

- Original discovery of a process change that may be an effect of a problem
- Documentation of the effects of the problem compared to *standard operating conditions*
- Follow-up with additional or amended effects on the problem
- Initial containment and control of the problem

Standard Operating Condition
A process variable set to produce the desired process output.

Without reliable execution of the above responsibilities, troubleshooting cannot succeed. One cannot prejudge which detail of an effect will be the key to the solution of a problem.

The shift process operator depends upon the day process operators for assistance and services. Sick and vacation relief are typical examples. Day personnel also maintain operating supplies and keep the necessary production records. Day personnel may very well be the designated liaison within the operations team concerning the plans and schedules for major maintenance activity.

1.6 PRODUCTION SUPERVISOR

The production supervisor is kept informed of any change in process conditions. He can approve a useful holding action until more about the process condition is known so that a rational correction can be worked out with operations team input. He considers what additional assistance is appropriate from mechanical and technical teams and makes the necessary contacts. Common practice is for the control room operator to assist the supervisor in his communications and to keep him informed of developments.

The production supervisor also prepares a summary report on shift activity that receives wide distribution throughout the operations team. The report is distributed to individual troubleshooters as a source of accurate process information. The production supervisor bases his report on information provided by the process operator on problems that need technical attention.

1.7 THE MECHANICAL TEAM

The mechanical team is composed of personnel responsible for the installation, calibration, repair, engineering specifications, and performance of all plant facilities. Buildings, piping, instruments, power generation, safety devices, and the availability of process equipment are major responsibilities of the mechanical team. The resources of the mechanical team are available to the process operator as troubleshooter on all problems that involve the care and performance of his equipment.

1.8 MECHANICAL-GENERAL

Mechanical-General personnel are part of the mechanical team and are responsible for the installation and repairs to process equipment and supporting facilities. Mechanical-General personnel service equipment routinely and are responsible for preventive maintenance programs. Preventive maintenance is an important phase of the overall troubleshooting effort that prevents unexpected equipment failure, and it deserves full support from the production team.

As members of the operations team, mechanical personnel contribute valuable input on equipment conditions that could hurt process performance if not corrected. Mechanical assistance is an invaluable troubleshooting resource in the analysis of equipment failure.

■ 1.9 MECHANICAL-ENGINEERING

Mechanical engineers are the experts on *prime movers* such as pumps and compressors. They provide an authoritative source of technical information on seals, gaskets, insulation, thermal stress, and the specifications on components used in piping systems. Mechanical engineers are also concerned with corrosion and materials of construction.

Prime Mover
A term that refers to pumps and compressors used to generate fluid flow.

In any problem that involves equipment malfunction and failure, the mechanical engineer is a key troubleshooter and consultant. It is the responsibility of the process operator to bring any evidence of substandard equipment performance to the immediate attention of the mechanical-engineering team.

■ 1.10 MECHANICAL-INSTRUMENT

Mechanical-instrument is a key arm of the mechanical team. Few technologies have been transformed in recent times as completely as instrumentation. Control systems, once predominantly pneumatic, are now electronic. The central control room and area workstations contain monitors to display process conditions and computer programs to process, store, and disseminate operations data. The troubleshooter today enjoys a wealth of detailed information made immediately available to the entire operation team as a tool to solve and prevent process problems.

Mechanical-instrument personnel are often assigned to specific plant operations in order to provide consistent coverage to control problems. Since the performance of the plant depends upon the reliability and accuracy of the process data that instruments provide, the process operator and mechanical-instrument personnel share information concerning the installation, calibration, repair, and performance of plant instruments.

It is not necessary for the instrument mechanic to be expert in process technology. As a team, the operator and instrument mechanic combine the knowledge of the basic control logic with knowledge of the process technology at work to accurately measure and record process variables.

Instrument engineers are responsible for the design of the equipment installed to measure and control process variables. When the process operator reports an instrument problem, he includes the instrument engineer as a key troubleshooter. Prompt discovery and correction of an instrument malfunction is often a contributing factor to the prevention of a process upset.

1.11 MECHANICAL-ELECTRICAL

Mechanical-electrical participation in the day-to-day operations of the plant occurs upon request by the production team. Electrical failures are infrequent because a great deal of design effort and investment is customary to provide highly reliable electrical facilities. An unexpected power outage is one incident that has the potential to produce major hazards in a petrochemical plant. If there is any evidence of an electrical problem, the process operator should not hesitate to have an electrician check it out.

The electrical engineer is one specialist that the process operator may not interact with on a regular basis. The electrical engineer devotes his troubleshooting effort to ensuring that all proposed process designs meet electrical-system reliability and safety standards.

1.12 THE TECHNICAL TEAM

It takes a number of applied technologies to effectively support the unit operations of a petrochemical plant. Process, instrument, electrical, mechanical, analytical laboratory, regulatory, and safety expertise are all essential technical support factors needed to maintain safe and productive operations within regulatory limits in today's industrial setting.

The production team relies upon the expertise of the technical team to provide a sound technical basis for standard operating conditions, standard operation procedures, chemical and physical changes, energy conservation, waste disposal, handling of hazardous materials, and all other technical aspects of plant operations.

For the process operator, technical-team specialists are a valuable source of process-technology expertise and are available to augment his technical knowledge of his facility. Most technical specialists enjoy talking about their work. The process operator should create opportunities to *network* with the technical specialists assigned to his area and fully exploit the technical expertise that each can provide on process technology. The

Network
A set of circumstances or activities that are dependent upon one another for success.

process operator is a valuable source of information that helps the technical specialist to improve plant operations. The plant operator, through his log and verbal exchanges, routinely keeps the technical team informed on plant operations under study.

Broad technical input on an abnormal process condition in order to reach a decision on corrective action that has a high probability of success justifies troubleshooting teams. The advice from technical experts on chemical and physical technology, instrument logic, and equipment performance is an invaluable asset that must be consulted before a process change is approved on a permanent basis.

The novice process operator should make mastery of the effective use of technical assistance a high priority during his on-the-job training. Success as a troubleshooter is dependent upon technical teamwork.

1.13 PROCESS CONTROL ENGINEER

The process control engineer works closely with production to meet safety, *yield*, *utility*, quality, and production goals. Systematic communication on process conditions between the process control engineer and the process operator is essential in meeting any performance goal. The process operator and the process control engineer are natural allies and team players with a common objective. Successful operation is the result of shared skill and knowledge of both that fully use the talents of the operations team.

Yield
The useful recovery of product from feedstock.

Utility
The hours available to produce product compared to total hours.

The process operator as troubleshooter seeks the advice and guidance of the process control engineer on any unexplained change in a process condition, no matter how small or insignificant the change may seem at the time. Few aspects of troubleshooting will do more toward the solution and prevention of process upsets than quick, consistent solicitation of troubleshooting assistance from the process control engineer.

The process control engineer and other specialists rely upon accurate, technical information from process operators to solve long-range process problems through teamwork. For example, the process control engineer studies the input of the process operator and makes a decision to bring in the outside assistance that is needed on a problem. The production supervisor is often consulted and, at the very least, is kept informed.

The importance of the free flow of information between technical and production teams cannot be overemphasized as a key process-operator responsibility and is in the best interests of a troubleshooter. Free exchange on the use of data on process conditions is critical to productive teamwork, and it is essential to troubleshooting as a tool to improve plant productivity.

1.14 PROCESS CONTROL LABORATORY

The process control laboratory is an analytical tool available to production and technical teams to monitor the specifications of process streams and finished products. There may be a resident chemist on hand to develop analytical procedures and to ensure the accuracy of results. He is available as the technical advisor on how to handle special samples.

The process operator takes most of the process samples. The care that he takes on each sample is an important factor in assuring the reliability of the analytical results.

Certification
A product conforms to specifications tested by approved analytical procedures.

A fairly recent trend is to have the process control laboratory results and the analytical procedures in use certified. Through *certification*, customers are assured that all phases of analysis conform to international standards.

1.15 CENTRAL ENGINEERING

Plants often have a central engineering group that provides basic mechanical, electrical, and instrument assistance throughout the plant. This is a practical strategy that increases the productivity of specialists and provides each production unit with access to a broad mix of technical talent. The downside is that the specific technical assistance needed may not be available at the precise time it is requested. Effective use of central engineering requires a special effort on the part of the process operator as troubleshooter to accurately document the effects of a problem in approved technical terms. Often there is a need to explain the technology at work in a specific unit. Precise technical detail is needed from all troubleshooters who work with central engineering.

In large plants, central engineering specialists are assigned to specific areas of the plant as part-time consultants for use by the operations team as needed. These specialists are expected to learn the basic technology of the process and to be familiar with the personnel assigned to the area. The process operator is of major assistance to central engineering by helping compare what is to what should be when defining a problem. The specific assignment of technical personnel to an area from any resource provides the process operator as troubleshooter with a sounding board on any process change when first observed, which translates into an effective troubleshooting and problem-prevention program.

1.16 SUMMARY

All petrochemical plant personnel are organized to assist the process operators in meeting common safety, production, quality, and cost goals. As Figure 1–1 shows, a number of individuals with diverse technical knowledge and skills are involved and must work in harmony as teams to succeed as troubleshooters.

The process operator with an education in process technology presents an opportunity to troubleshoot process problems. As a productive member of troubleshooting teams, he joins the technical experts in place. As you study the process operator role in effective troubleshooting, use this section as a reference to select the assistance needed on assigned exercises.

CLASS EXERCISES

The following class exercises—1.1, 1.2, 1.3, and 1.4—are provided to improve communication skills, practice team troubleshooting, and to promote understanding of the section material. Each team selects members from Figure 1–1 that are needed to solve the problem and work together on a solution.

The exercises can be approached in several ways:

1. The entire class functions as a team and solves the problem in the exercise.
2. Each class member works out a solution, and then combines his input with the solutions of others on a team selected by the instructor.
3. The instructor selects the team members and appoints one as team leader. This works well in classes up to twenty students.

EXERCISE 1.1 REACTOR CLEANUP

The instructor selects the team members and appoints one to be team leader. Individual jobs are selected by each team.

Team Members	Team Leader
______________________	______________________
______________________	
______________________	
______________________	

The team plans the cleanup of a reaction system, which contains toxic materials. Gross deinventoried material is recycled, but the final purge of material is sent to a waste-treatment facility in a remote location.

Instructions

1. Choose the team jobs from Figure 1–1 with the skills needed to do this job safely.
2. The team leader needs technical or operations input from each member, which means that the knowledge and skills on the team are chosen to plan a successful cleanup.
3. The team prepares the plan to clean up the reactor. A sketch and time line help to organize the discussion.
4. The team presents its plan to the class, with each team member describing his or her input.

5. The class discusses the viability of the team plan.
6. Each team member prepares a written report on his participation in the plan for the instructor.

Time (flexible)

Instructions 1 and 2	15 minutes
Instruction 3	15 minutes
Instructions 4 and 5	10 minutes

Note: The instructor performs as the process control engineer.

EXERCISE 1.2 PROCESS SPILL

The instructor selects the team members and appoints one to be a team leader. Individual jobs are selected by each team.

Team Members	Team Leader
______________________	______________________
______________________	
______________________	
______________________	

The team plans the cleanup of a large, liquid spill on the ground floor of a process building that contains toxic chemicals. A small quantity of the spill evaporates due to a significant vapor pressure. The bulk of the spill drains naturally via an open trench to a covered building sump located 50 feet away.

Instructions

1. Choose the team members from Figure 1–1 with the skills to plan this cleanup safely.
2. The team discusses the input needed from each member and concurs that the knowledge and skills on the team can indeed plan a successful cleanup.
3. The team prepares a plan to clean up the spill with input from each member. A sketch and time line help in guiding the discussion.

4. The team presents its solution to the class, with each member describing his or her input.
5. The class discusses the viability of the team plan.
6. Each team member presents to the instructor a written report on his or her participation.

Time (flexible)

Instructions 1 and 2	15 minutes
Instruction 3	15 minutes
Instructions 4 and 5	10 minutes

Note: The instructor is the process control engineer.

■ EXERCISE 1.3 TANK ENTRY

The instructor selects the team members and a team leader. Individual jobs are selected by each team.

Team Members	Team Leader
________________	________________
________________	
________________	
________________	

The team plans the entry into a 500,000-gallon tank in a mixture of liquid wastes service. The tank is due for a complete internal, metallurgical inspection, which will take 8 hours.

Instructions

1. Choose the team members from Figure 1–1 with the knowledge and skills to do the job safely.
2. The team discusses the input needed from each member and concurs that the skills on the team can indeed plan a successful entry.
3. The team prepares the plan to enter the tank with input from each member. A sketch and a time line help to organize the discussion.
4. The team presents its plan to the class with input from each member.

5. The class discusses the viability of the plan.
6. Each team member presents to the instructor a written report on his or her participation.

Time (flexible)

Instructions 1 and 2	15 minutes
Instruction 3	15 minutes
Instructions 4 and 5	15 minutes

Note: The instructor is the process control engineer.

EXERCISE 1.4 DE-INVENTORY AN EXTRACTION TOWER

The instructor selects the team members and appoints one to be team leader. Individual jobs are selected by each team.

Team Members **Team Leader**

____________________ ____________________

____

____

____

The team plans the de-inventory and cleanup of an extraction tower that uses water as the solvent. Water is also the dense phase in the extraction process. The raffinate is a flammable material and is naturally immiscible with the solvent. The solute is a mildly toxic salt. Extract is normally shipped in tank trucks offsite for disposal. Purge material is sent to a plant decanter as a precaution, then on to aqueous waste.

Instructions

1. Choose the team members from Figure 1–1 with the skills to plan the job safely.
2. The team discusses the input needed from each member and concur that the knowledge and skills on the team can indeed plan a successful de-inventory.

3. The team prepares a plan to de-inventory the extraction tower. A sketch and a time line help guide the discussion.
4. The team presents the plan to the class with input from each class member.
5. The class discusses the viability of the plan.
6. Each team member presents to the instructor a written report on his or her participation.

Time (flexible)

Instructions 1 and 2	15 minutes
Instruction 3	15 minutes
Instructions 4 and 5	15 minutes

Note: The instructor is the process control engineer.

CHAPTER

2

Communications for Troubleshooters

LEARNING OBJECTIVES

- Understand the inherent difficulties in communications among troubleshooting teams and how the process operator overcomes them.
- Explore the role that the process operator fills as a key communicator, which sets the stage for teamwork in troubleshooting.
- Practice verbal and written communications in troubleshooting situations.

KEY TERMS

Process conditions
Communications network
Process upset
Technical specialists
Discovery
Workstation
Log
Data sheet
Shift log
Systemic
Shift team
Relief team
Effect
Procrastination
Intrinsic problem
Analysis
Corrective action
Verbal communications
Intercom
Sketch

Process Condition
A process variable that is controlled to achieve steady-state operation that produces a predicted chemical and/or physical change.

Communications Network
An agreement on the exchange of information among those who need to know.

Process Upset
An operation unexpectedly carried outside standard operating conditions.

Technical Specialist
An expert in one or more aspects of process technology.

Discovery
The first step in the problem-solving process. Sort out effects of a problem into a logical order.

Workstation
A computer designed to carry out a specific agenda of assistance to operations.

Log
A written account of plant activity on a planned schedule.

Data Sheet
A form designed for entry of specific process conditions.

Shift Log
A continuous record of plant operations.

Systemic
Pertaining to a condition that consistently exists in a given system.

This chapter examines typical plant-communication needs and the inherent problems the process operator as troubleshooter faces when trying to meet those needs effectively.

Reliable distribution of accurate technical data on *process conditions* throughout an industrial organization relies upon a well-designed *communications network*. Communication of accurate and timely effects that permit correction of a condition with the potential to cause a *process upset* is at the heart of productive troubleshooting. Installation of adequate equipment needed to communicate among the troubleshooting teams is easy to come by. However, dedicated reporters of timely and accurate information are the essential requirement of the successful communication network and are far more difficult to establish.

In a successful network, the facilities in place to maintain effective communications between shift personnel and the supporting *technical specialists* are reviewed continuously. The type of information to be communicated, audience, reciprocity, and procedures needed to communicate effectively are studied for adequacy. Any reduction in the time between *discovery* of a plant problem and communication to solicit troubleshooting assistance is a solid troubleshooting accomplishment. Using the computer *workstation* to record and transmit technical information on process conditions between day and shift teams is an effective communications tool for the process operator as troubleshooter.

The process operator is at the hub of the communications network that feeds production information to the operations team. His or her constant presence in the plant is ideal for detecting and reporting technical details on plant conditions. The process operator, who is educated in the chemical and physical changes that make a chemical process work, is an effective communicator within the troubleshooting team.

The written communications, including *logs* and *data sheets*, that the process operator prepares and uses in his or her daily work contain essential information that all troubleshooters rely upon. Few communications are more important than keeping *shift logs*, process data sheets, and laboratory reports current and accurate.

2.1 THE ABSENCE OF PERSONAL CONTACT

The process operator has limited opportunity on shift to speak person-to-person with the technical and mechanical specialists. The probability that a problem will occur at night, on a holiday, or during the weekend is much greater than on Monday through Friday between eight to five, which are the standard working hours for specialists. Unless that *systemic* deterrent in communications between day and *shift teams* is overcome, team troubleshooting simply cannot function at acceptable maximum efficiency.

An organized exchange of accurate information between day and shift personnel on the production team is essential to the troubleshooting effort. To overcome the absence of personal contact, the process operator must see him or herself as a member of two teams that communicate in one voice (Figure 2–1). The first team is the shift of five or more field process operators that run a specific, continuous process with four or more major steps from feed to end product. This team includes a central control room (CCR) operator and is supported by a production supervisor. All the process operators on the team work with standard operating procedures to keep each unit safely within standard operating conditions. Most communications are intended to

Shift Team
An organized group of process operators needed to run a multi-unit, petrochemical process.

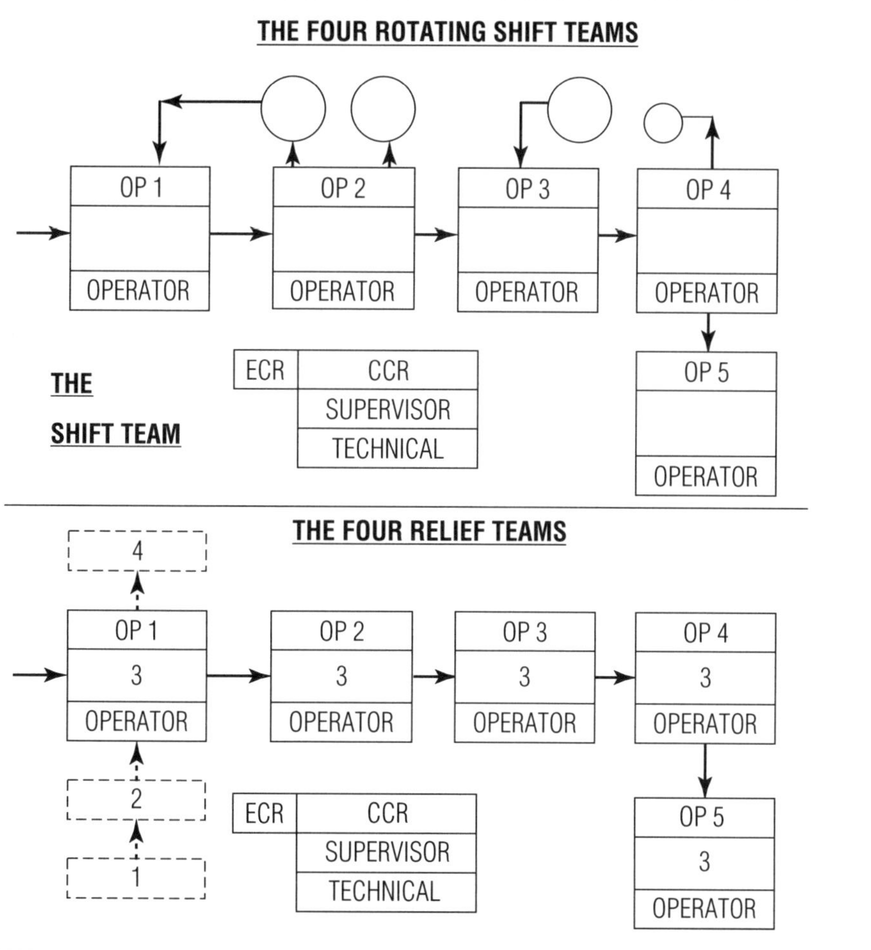

Figure 2–1
The Shift and Relief Teams

sustain continuous, quality production. The production supervisor is the designated communicator with the technical specialists (in the event their assistance is required). The activity reported by each process operator is collected into a cohesive shift log for broad distribution by the shift supervisor.

Relief Team
Process operators scheduled to provide relief and work the next shift.

The *relief team* includes the four process operators who are responsible for the performance of one of the several steps in the continuous process. These four, collectively, are the source of the immediate and detailed information on process conditions that troubleshooting teams rely upon to do their jobs as troubleshooters. Relief teams support the troubleshooting effort as the primary source of continuous, timely, and accurate information on conditions in a specific unit.

The following steps will assist the shift operator in assuming his or her responsibility as a key communicator in the troubleshooting effort:

Effect
A measured, observed, or sensed abnormality in the interaction of people, places, and/or things.

Procrastination
Delay of a necessary activity until a future time.

Intrinsic Problem
A condition that repeats in a continuous process that detracts from performance.

Analysis
The deductive reasoning phase of the problem-solving process. The study of the effects of a problem with the objective to disclose all probable causes.

Corrective Action
A change that is made to minimize the effects of a process upset and that is aimed at eliminating root cause.

1. Learn the functions and identity of fellow troubleshooters within the organization and keep current on changes in job assignments. The process operator must know the correct team member to contact when he or she needs troubleshooting assistance.
2. Work through the shift supervisor to communicate with the mechanical and technical specialists during their off hours. Technical responsibility for continuous coverage of plant units is assigned to knowledgeable individuals at all times.
3. Upon discovery of a problem, communicate its *effects* quickly. *Procrastination* can be a serious set back in effective troubleshooting, making that important chance to prevent a process upset impossible.
4. Consistently follow up to check the results of action taken to correct a problem.
5. Stay abreast of the status of the continuous effort to troubleshoot *intrinsic problems*. Discovering a change in effects may be the information needed to determine cause.
6. Provide answers to the questions of *where*, *what*, *when*, and *how* as a means to keep communications organized and tied to a specific problem.
7. Document all assistance received from others and identify each participant involved in a problem from the time of discovery, through *analysis*, to the *corrective action* taken. Each person involved in a problem is a potential source of unique information that assists the troubleshooting effort.

2.2 VERBAL COMMUNICATIONS AT SHIFT CHANGE

Verbal Communications
Communication by word of mouth.

The process operator sees in shift change a crucial responsibility to take stock of all activity in his or her area. Few troubleshooting activities are more successful in problem prevention than carefully organized *verbal communications* designed to send the next shift operator off to a good start with his or her priorities set on action needed on plant activities.

The verbal exchange between a process operator and his or her relief at shift change emphasizes priorities and establishes a sense of urgency on pressing problems. Only the process operator being relieved can highlight the problems at hand and refer to additional sources for detail, such as the shift log. Shift change is also a proper time to cover work in progress that deserves immediate and close attention. In summary, the process operator instructs his or her relief on tasks that need tending to early in the shift and suggests specific ways to do so.

2.3 THE SHIFT LOG

Verbal exchange is not a substitute for a well-written shift log. The quantity of verbal information (particularly, the amount of technical detail) that a person can retain in a verbal exchange is limited. The process operator as troubleshooter does not rely upon memory as a reliable source of technical detail. All plant personnel look to the written log for detailed information on subjects covered during the verbal exchange.

Another objective of the shift log is to record unit performance from shift to shift even if no crisis on a problem exists. Information on work in progress, work completed, preparations for future work, priorities on all activity, and reminders on safety concerns are usefully and thoroughly found in the shift log.

As discussed under the relief team, a log on the performance of parts of a continuous operation that is covered by one job assignment is shared equally by four process operators. They provide specific information to the operations team on problems that require technical and mechanical assistance on their part of the process.

The quality of the content of the shift log deserves a high priority, as it is a communication tool used by all the troubleshooters seeking hard information. There is no comparable document in which to record progress (or failure) of production effort, to document the status of process activity, or to request follow-up on an action taken earlier.

The process operator as troubleshooter writes the shift log to fill four unique communication goals:

1. Provide an accurate record of technical data for use in the reconstruction of the effects of a problem within an accurate time frame.
2. Provide new insight and make corrections to past logs upon discovery of additional information. Entries in the current log with a reference to past logs meet this goal. *Past logs should never be altered.*

3. Make information available to all parties with a need to do their job.
4. Provide a continuous record of shift activity that becomes a permanent reference and a history of unit performance.

To communicate efficiently in the shift log, the process operator as troubleshooter organizes his or her entries by consolidating all information about one problem in one location. The format and content of information in the log can be targeted for the convenience of the intended user of the information. New information on active problems is referenced to past logs for quick study by the operations team. This simple move helps a continuity of communications within the troubleshooting team.

New problems deserve special effort; the log should provide thorough background data and the technical detail needed to get fellow troubleshooters on board immediately. Bringing the right technical attention armed with the right technical data to a problem quickly is the goal of all effective troubleshooters.

A pocket notebook, used to record effects as they are discovered, provides invaluable recall in preparing a shift log. A complete shift log and verbal exchange on shift activity is made much easier by notes that contain the numerical detail needed to describe events precisely and accurately.

To gather information for a meaningful shift log, use the following steps:

1. Take notes on effects as soon as possible after discovery of the problem and the action taken to normalize process conditions.
2. Concentrate on accuracy in the numerical data.
3. Answer the questions *what*, *where*, *how*, and *when* for each discovery.
4. Note all personnel involved, the action taken, and the results achieved.
5. Follow up and communicate any corrections or additional information.

2.4 THE INTERCOM

Intercom
A speaker service between the central control room and the operators in process units.

The *intercom* (radios are often used) is the system most often provided for continuous communications between members of a shift team. Intercom communications between the central control room (CCR) and each field operator are used primarily to exchange information that is needed to synchronize activity within the shift team.

The intercom can be made available to bring technical expertise in on an unexpected problem immediately. This quick communication on the early discovery of an unexpected process condition is a classic troubleshooting strategy to prevent serious process upsets.

Communications about an unexpected problem usually follow the general path shown in Figure 2–2. The field operator discovers a change in his or her operation and reports the initial effects to the CCR. The CCR operator alerts the production supervisor and informs personnel in other plant units that they may be affected and to be prepared. Suggestions from the CCR operator and the supervisor on action to be taken are normal assistance that the field operator expects and needs. The first-line troubleshooting team is at work.

The field process operator continues to gather information and follows up with the results of any corrective action taken to contain the problem, keeping the CCR operator fully informed on results. If necessary, the production supervisor contacts technical specialists who can be of assistance. In a minimum amount of time, an organized troubleshooting team effort from the operations team is in place to analyze the effects of the problem based on the information supplied by the field process operator.

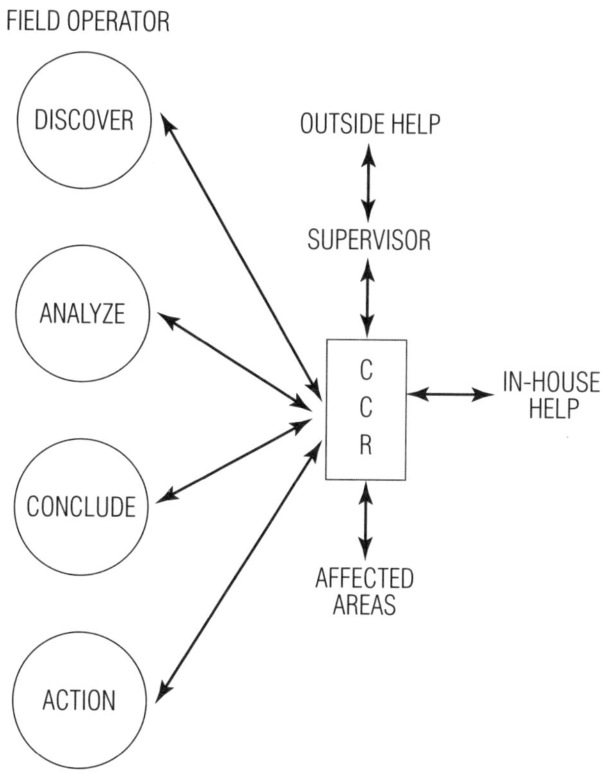

Figure 2–2 Intercom Communications During an Unexpected Problem

As problem analysis proceeds, preliminary conclusions on the probable cause are reached with input from the troubleshooting team. The troubleshooting team agrees on the most probable cause and recommends the appropriate corrective action that will eliminate the problem. The process operator implements the corrective action and keeps the team informed on the results.

The above scenario is somewhat idealistic with its expectations of perfect communications and perfect solutions. However, if rigorously applied, the troubleshooting team will often experience good short-term results in the control or even the prevention of a process upset.

The work of the process operator as troubleshooter is not complete when the problem is under control. All pertinent information, communications, and action taken on the problem are collected and documented in the shift log. At this point, followup is needed to provide assurance that the root cause has been identified and eliminated.

Note: A tape of intercom conversations can be used to reconstruct problems and become a helpful training tool.

2.5 THE IMPORTANCE OF TERMINOLOGY IN COMMUNICATIONS

The study of communications in petrochemical plants includes a great deal of emphasis on learning and using correct technical terms and symbols. This is the important knowledge that the technically educated process operator can bring to his or her on-the-job training. He or she is able to read and better understand diagrams and process descriptions that are used in the classroom to depict the facilities in the field.

Adding to one's technical vocabulary and using correct terminology in all communications is a skill both admired and needed in a troubleshooter.

2.6 GENERAL COMMUNICATIONS

Shift process operators have a limited opportunity to attend meetings with day personnel, where information is exchanged on safety and plant operations in general. The process operator who can attend these meetings adds to communications within the production team—an opportunity is there to do some positive troubleshooting—by sharing information on the status of process problems in his or her area.

Technical and mechanical specialists assigned to a process unit pass through frequently and often contact the process operator in search of answers to their questions on a specific process problem they are working on. The process operator as troubleshooter takes the opportunity to answer their questions from his or her viewpoint and to exchange information on a broad range of area problems. This one-on-one opportunity to review information in shift logs allows the process operator to question information in technical reports written by the specialists.

A good troubleshooter approaches all visitors to his or her area and inquires after their mission with an offer of assistance. This is an opportunity for the process operator to stay abreast of the personnel in the organization and their responsibilities as troubleshooters. Checking out visitors is also a good safety practice. The process operator confirms the visitor is equipped with the required safety apparel, points out areas of restricted access, and identifies escape routes.

A major operations team effort goes into the use of safety rules and work practices to eliminate unsafe practices that result in injury. The process operator as troubleshooter is in a position to comment on the changes needed to safety rules and work practices which keep safe operations a foremost troubleshooting objective.

2.7 COMMUNICATIONS WITH KEY PERSONNEL

The process control engineer needs to be the best-informed specialist. To fill the troubleshooting role, he or she must stay abreast on all aspects of the operation and be available as a technical consultant to the operations team. By working closely with the process control engineer, the process operator adds his or her contribution to the solution and prevention of intrinsic process problems.

The process operator works with the process control engineer in a continuous review of standard operating procedures and conditions for areas of improvement. As previously stated, the process operator and the process control engineer are natural allies and team players who share the responsibility to obtain from the operations team all technical assistance needed.

The process control engineer and other specialists use energy and material balances as major technical tools to analyze process performance. Flow rate, temperature, pressure, and stream composition are typical process conditions; the process operator is in the ideal position to supply these data and to ensure accuracy. The process operator as troubleshooter communicates with the intent to keep all troubleshooters supplied with the latest information. He or she does not judge who will find the information valuable.

The process operator as troubleshooter thinks in numbers, talks in numbers, and writes in numbers for the benefit of those who solve problems through the use of numbers—this includes every technical specialist on the plant. The process operator can be overwhelmed by numbers, unless he or she masters good techniques in handling technical data:

- During the formal on-the-job training period, do some homework. Review the events of each day and write down your questions. Concentrate on phrasing questions using the correct terminology. Bring them up the next workday at the first opportunity. If possible, make a sketch of facilities, which helps to clarify the question. This excellent habit can be put to use every day as a troubleshooter.
- As a communication tool, there is no substitute for good notes and *sketches* made from those notes. Add as much numerical data to the sketch as you can. The ability to communicate through sketches with numerical supplements is an effective way to achieve agreement about technical detail. There is no better way to clarify or expose gaps in technical information.

Sketch
A drawing without scale used to show a particular part of a process.

2.8 MEMOS, REPORTS, AND VERBAL EXCHANGE

Technical specialists, plant staff, and other support groups are members of troubleshooting teams that work on intrinsic problems. Frequent status reports cover technical progress and make recommendations on future work. These status reports deal with some aspect of the relationship between people, places, and things in plant operations that need improvement. The process operator as troubleshooter reads reports on the progress on intrinsic problems in his or her area—and often sees a way to contribute to a solution with new information and suggestions.

The troubleshooter reads incident and injury reports regardless of source and location. The description, discussion, and recommendations made in reports to prevent the recurrence of the unsafe event adds to the knowledge and skill of the process operator in the handling and prevention of similar incidents.

The motivated troubleshooter networks with key personnel, exchanging information and sharing experiences, which help both do their jobs better. This feature of teamwork improves troubleshooting performance throughout the operations team.

2.9 SUMMARY

Effective communication of the effects on process problems, both verbally and in writing, is a key purpose of verbal exchange and the shift log. Learning to communicate process conditions effectively on the job is an important feature of the troubleshooting needed to run a process safely and consistently, since troubleshooting relies upon a well-informed operations team.

There is considerable opportunity to practice both written and verbal communication skills in the chapters ahead.

CLASS EXERCISES

The three exercises in this chapter, Class Exercises 2.1, 2.2, and 2.3, are intended to demonstrate the communication skills needed among troubleshooters when working as a team.

The instructor chooses teams of three to five students and designates a team leader. An oral presentation (verbal exchange) by each team on its findings as well as writing a "shift log" are good ways to practice communication skills. The entire class is asked to critique the oral presentation of each team. A group discussion on how to attack a plant problem led by those who are close to it is a common communication forum at a petrochemical plant. The purpose is to bring as much knowledge and skill as possible on specific problems to the table.

EXERCISE 2.1 LEAK ON A FLANGED CONNECTION

Team Members	**Team Leader**
______________________	______________________
______________________	
______________________	
______________________	

Read the introduction to the exercise "*Leak on a Flanged Connection,*" Chapter 8.

Instructions

1. Choose the appropriate jobs from Figure 1–1 for the troubleshooting team.
2. Establish a communications network within the team. What lines of communication are needed? (See Figure 2–2.)
3. Draw a diagram of the communications network and describe how it works.
4. Present 1, 2, and 3 to the class.
5. Prepare a written summary for the instructor in the form of an entry in a shift log.

Time

Instruction 1	10 minutes
Instruction 2	15 minutes
Instruction 3	15 minutes
Instruction 4	As necessary
Instruction 5	Provided by instructor

Note: The instructor will provide technical assistance as requested.

EXERCISE 2.2 THE REFRIGERATION UNIT

Team Members **Team Leader**

________________________ ________________________

____

____

____

Read the introduction and the situations A, B, and C for the exercise, "*The Refrigeration Unit,*" Chapter 8.

Instructions

1. Choose the appropriate jobs for the troubleshooting team.
2. Establish a communications network within the team. What lines of communication are needed? (See Figure 2–2.)
3. Draw a diagram of the communication network and describe how it works.
4. Present 1, 2, and 3 to the class.
5. Prepare a written summary for the instructor as an entry in the shift log.

Time

Instruction 1	10 minutes
Instruction 2	15 minutes
Instruction 3	15 minutes
Instruction 4	As necessary
Instruction 5	Provided by instructor

Note: The instructor will provide technical assistance as requested.

EXERCISE 2.3 WASTE DISPOSAL

Team Members	Team Leader
____________________	____________________
____________________	
____________________	
____________________	

A toxic waste tank will be full in 48 hours. The onsite waste disposal facility for material from this tank is available 50 percent of the time on a first-come-first-served basis with 24 hours notice. An offsite contractor will dispose of the material but insists on a definite ship date with a 48-hour lead time.

Instructions

1. Choose the appropriate jobs for the troubleshooting team.
2. Establish a communications network within the team. What lines of communication are needed?
3. Draw a diagram of the communication network and describe how it works.
4. Present 1, 2, and 3 to the class.
5. Prepare a written summary for the instructor as an entry in the shift log.

Time

Instruction 1	10 minutes
Instruction 2	15 minutes
Instruction 3	15 minutes
Instruction 4	As necessary
Instruction 5	Provided by instructor

Note: The instructor will provide technical assistance as requested.

CHAPTER

3

Detection and Management of Problems

LEARNING OBJECTIVES

- Understand the differences between intrinsic and unexpected problems.
- Accept the responsibility to detect and manage problems as a process operator as troubleshooter.
- Accept good communications by the process operator as essential to the management of problems.
- Understand how effective plant patrol is based on the detection and management of problems.
- Realize that the detection and management of problems is an operations team charter, and the process operator fills a key position.

KEY TERMS

Detection
Management
Effect
Productivity
Intrinsic problem
Progressive effects
Dead time
Cause
Checklist
Timetable
Unexpected problem

Detection
The initial discovery and communication on the effects of a problem.

Management
The act of organizing or controlling an activity at an acceptable level.

Effect
A measured, observed, or sensed abnormality in the interaction of people, places, and/or things.

Productivity
A comparison of process performance to expectations, including utility and production rate.

Intrinsic Problem
A cyclic activity that repeats in a continuous process, detracting from performance.

Progressive Effects
One effect becomes cause and creates a second effect, etc.

Dead Time
Wasted or unused time in an otherwise full schedule of planned activity.

Cause
The improper interaction among people, places, and things that results in a process problem, which is detected through abnormal effects.

The process operator is positioned ideally to lead the operations team in problem *detection* and to contribute as a team member in problem *management*. His observation skills are the prime source in early detection of a process change from noticeable *effects* in an ever-changing work environment.

The industrial workplace is under continuous change that involves the interaction of people, places, and things. The early detection and management of unexpected change in this interrelated mix of activity does have a positive impact on safety and *productivity*. Early detection of any deviation from normal is a much-needed skill that the successful process operator as troubleshooter continuously puts to use.

3.1 THE INTRINSIC PROBLEM AND THE TROUBLESHOOTER

In the broadest of terms, an *intrinsic problem* is an undesirable process condition that repeats and must be managed properly to minimize loss in plant utility. Examples among many possibilities are lines that plug, connections that leak, scale buildup on heat transfer surfaces, faulty instruments, worn-out equipment, and human error.

A number of processes house intrinsic problems that are managed through close surveillance. An intrinsic problem displays *progressive effects* that are monitored routinely by the skilled troubleshooter. The troubleshooter extrapolates the data on effects he collects on an intrinsic problem to a point in time where the need for corrective action can be forecast. This is often accomplished on a plot of performance versus time. This plot, which tracks problem development, permits development of a timely plan on corrective action. Predicting the time corrective action is needed leads to advance scheduling of the needed resources, which boost productivity by a reduction in *dead time*. A successful ploy in the detection of a process change is to compare what is to what should be during plant patrol. Also, an excellent attitude in troubleshooting is to continuously expect change as an inherent feature of any dynamic activity.

The knowledgeable troubleshooter accepts the premise that any change in a process variable or in equipment performance that deviates from an accepted standard has a logical *cause*. The experienced troubleshooter expects any deviation from normal to worsen unless the cause is detected and corrected. It follows that any detected process change deserves an explanation. The initial action by the process operator includes the documentation and communication of what, where, when, and how the abnormal effect was detected, which is information needed for use by troubleshooters in the operations team to solve and eliminate the problem through exposure of the root cause.

Knowledge of process technology is essential as a basis to analyze effects of a problem with the goal to determine the cause. A process operator, who is knowledgeable in process technology, is prepared to communicate a process change using correct terms and symbols. He or she can supply the basic data on effects in the technical terms that communicate clearly to the technical specialists. The operations team, including the process operator, will solve process problems when they are provided with timely and accurate information to work with.

Continuous growth in the knowledge and skill of the process operators to detect early effects of a problem with people, places, and things is a significant step toward a truly effective troubleshooting program. Exposure of the cause of a problem frequently produces a needed change in a standard operating procedure, an improved design of equipment, and a higher level of logic in instrumentation.

One major obstacle to troubleshooting intrinsic problems is the absence of conviction by the process operator that progress toward a solution is being made or that a permanent solution is a realistic goal. Managing the intrinsic problem becomes a part of his daily routine, which the process operator has learned to accept as an inherent part of his job. However, the idea that working to better manage intrinsic problems will lead to continuous improvement in process performance is embraced by all who would be effective troubleshooters.

The dedicated troubleshooter is never complacent. He understands that the level of productivity reached today will not be competitive tomorrow. A review of the history on incremental progress to minimize the effects of an intrinsic problem reveals, invariably, that continuous troubleshooting by knowledgeable teams is a proven tactic to improve plant performance.

3.2 MANAGING INTRINSIC PROBLEMS

Standard operating conditions are set with control limits on the process variables that produce the desired result in a petrochemical process. These conditions create intrinsic problems that must be managed in order to meet acceptable performance goals.

Plugged lines and fouled heat transfer surfaces are commonplace in some processes. Few chemical compounds react ideally to chemical and physical change. Catalysts lose activity and ion-exchange resins lose capacity over time. Process material can build up on metal surfaces. Operating conditions may produce corrosion as a result of elevated temperature and pressure. These factors and many more are typical operating requirements that test the limits of materials of construction; furthermore, corrective maintenance causes loss of plant utility. Whoever manages these losses most effectively has the advantage over competition.

Intrinsic problems can be difficult to eliminate permanently. They usually have well-disguised effects from technically complicated causes. The intrinsic problem, with well-hidden causes and effects, is the prime target for improvement by the process operator and fellow troubleshooters. New data on process technology makes the intrinsic problem more manageable. When the effects on intrinsic problems are closely monitored, the corrective action necessary to limit a loss in productivity can be scheduled with considerable precision. The process operator has become a troubleshooter when he fills this need.

For intrinsic problems that involve process technology, technical teamwork is a key feature of the troubleshooting effort. The process operator continuously updates the progressive effects on an intrinsic problem, which keeps the technical experts working with the latest data toward a solution. Incremental and continuous improvement in process performance will be made, and the eventual elimination of the problem stays a distinct possibility.

3.3 TEAMWORK IN PROBLEM DETECTION

The central control room (CCR) operator alerts the field operator of an unexpected change in a process condition that has surfaced through a change in an instrument readout. This two-man troubleshooting team may or may not be on to an early effect of what could be the development of a major process upset. The field operator teams up with the CCR operator to analyze the change and to determine what impact, if any, further development could have on normal operation. Is corrective action needed immediately, and in what form? This example of troubleshooting is a daily occurrence, which explains the value of technically educated troubleshooters in every job at all levels, and at all times in coming up with the right decision on action to manage a process change with acceptable risk.

The process operator as troubleshooter documents effects, records action taken, and who participated in that action. This could be the information needed by a technical specialist to advance her work toward the solution of an intrinsic problem. Once again, the pocket notebook is very handy to record the facts for recall later in the shift log and the verbal exchange.

An abnormal condition may be brought to the attention of the process operator by any member of the operations team. The informer could be a mechanic, a fellow operator, or a technical specialist in the area. The process operator as troubleshooter welcomes this dimension of teamwork as an extension of his own observations on plant conditions. The member's sharing his or her information with the process operator immediately upon detection of an abnormal process condition demonstrates the essence of team work in troubleshooting.

The process operator puts to use all the information he is given to better prevent process upsets and manage problems. This dimension of teamwork is encouraged by followup with the informer on any action taken. What better way can one show appreciation for a fellow team player?

Consistent detection and management of process problems is not an accomplishment of chance or good fortune. There is process knowledge and skill along with hard work being put to use. Effective problem-detection and -management programs, which detect faults in technology or equipment, are carefully organized around safety and housekeeping inspections, a universal communications network, and technical teams who are immediately informed on any notable process change by the process operator.

3.4 THE PLANT PATROL

The process operator gathers a large percentage of the information on process conditions that troubleshooters need through plant patrol. The plant patrol is a highly productive troubleshooting tool in the hands of a well-trained and technically knowledgeable process operator. As a means to detect problems early on in a continuous process, there is no substitute for consistent inspection by a qualified troubleshooter. For the plant patrol to become an invaluable troubleshooting tool, exceptional perseverance is required in a work environment where days may pass without a serious deviation from normal process conditions.

In practice, plant patrol is a thorough comparison of what is to what should be in process conditions, equipment performance, and human activity. Any observed deviation from normal initiates a troubleshooting effort to find cause. Notes are taken for documentation in the shift log. The well-organized troubleshooter may use a *checklist* of hot spots prepared from information passed on during verbal exchange and by reading the shift log.

Checklist
A list used as a reminder of tasks one is responsible for completing.

A productive plant patrol with effective troubleshooting as the goal can be defined in specific steps that fully use operating skills and technical knowledge.

1. Read the shift log and compare the content of the written report with the verbal exchange received at shift change.
   - Be sure to include past logs that cover time off, which instills a feel for the general flow of activity, past and present.
   - Make a *timetable* (checklist) of specific tasks to be completed.
   - Follow up on information that does not clearly define what action is needed.

Timetable
A checklist organized against time. Activities are set down from first to last.

2. Review current and future activity with the process operators upstream and downstream. In a continuous process, this is essential communications that further teamwork.
   - Discuss planned activity with the CCR operator and the shift supervisor.
   - Ask the CCR operator and shift supervisor to communicate a timetable on shift activity to others who need to know within the operations team.
3. Check on all work in progress. Review the status and confirm the expected completion time with the worker(s).
   - Solicit or ask questions about safety and work practices to ensure that these important points have been covered.
   - Make clear to all workers that you are available to answer their questions as the work proceeds, and that you wish to be kept informed of any change in work status.
   - Review the lines of communication and workplace standards that everyone is expected to follow.
4. Inspect all work reported as completed on the previous shift. Share any concerns with the shift supervisor for immediate action. This is a sound troubleshooting tactic to prevent an *unexpected problem* at a time when taking corrective action would be far more difficult.

Unexpected Problem
A change where normal work practices did not detect the effects.

5. Check on sites where future work is scheduled.
   - Plan the preparations for future activity that are your responsibility in a timetable that eliminates conflicts for your time.
   - Look ahead at activity through at least the next shift with the intent to prepare your relief for an efficient start of his tour. This is a deliberate effort to work as a team and pays huge dividends in not allowing shift change to create an interruption in any scheduled activity.
6. Review laboratory results, memos, and technical reports that involve the operations team.
   - Give special attention to any revision to an operating condition or procedure. Carefully review all incident reports for valuable information on problem detection, management, and prevention.
   - If there are any questions, make an opportunity to check with the source of the information.
7. Select a specific area or unit operation for a detailed inspection. This close inspection is excellent insurance that prepares the troubleshooter to manage that infrequent process change that requires quick action to prevent a process upset.
   - Rotate the inspection to cover all facilities over a reasonable period of time.
   - A complete inspection includes a mental review of the function of every line, valve, instrument, and equipment piece.

  - As a useful refresher, mentally apply the standard operating and safety procedures set up to cover an emergency.
  - Note and report any detected abnormality.
  - Keep your notes as background information to review before starting the next inspection.
8. As the exterior of a facility is being inspected, use knowledge of process technology at work to mentally follow the physical and chemical changes that are being carried out inside. Do not hesitate to communicate with the technical specialists on any new or unusual change in an expected condition.
9. Give a facility with a history of problems a more frequent inspection. Update the status of all intrinsic problems and confirm the timing to schedule corrective action.
10. Expect the unexpected.
    - Be alert for any subtle change in a process condition.
    - Bring an unusual effect to the attention of the shift team. Procrastination prevents effective problem prevention through prompt troubleshooting. The CCR operator and the shift supervisor can bring in technical and mechanical assistance to help with a technical explanation.
11. Make every plant patrol a safety and housekeeping inspection.
    - Refuse to compromise on established plant standards and procedures.
    - Encourage all workers to use work practices, which are clearly explained and consistently enforced with the intent to eliminate hazards and maintain workplace housekeeping.

The eleven suggestions above will result in an effective and rewarding plant patrol in the detection and management of problems. Communicate and document the plant-patrol findings in the shift log for use by all the operations-team members.

3.5 THE HUMAN FACTOR

Even well-trained personnel, with the best of intentions to do their jobs, make mistakes. A troubleshooter must deal with occasional physical impairment, negligence, poor communications, inadequate training, and poor workmanship.

Many problems that involve human error are the result of a breakdown in communications. A task is easily mishandled when incorrect information and inadequate instructions are provided. Another potential source of human error is the inexperienced worker who needs assistance to handle the job but is reluctant to ask for it. A problem in communications is common where large numbers of workers and their activities must be closely coordinated to prevent conflicts. The process operator as troubleshooter successfully detects and manages problems founded in human error by the simple but highly consistent practice of close and consistent followup on all activities in his area.

Procedures designed to successfully manage the human factor in troubleshooting are:

1. Keep current on work schedules and work orders.
2. Be a close observer of all activity in the area. Pose questions to workers on shared responsibility and expect knowledgeable answers. Safety and housekeeping standards among all parties involved are routinely checked and enforced.
3. Communicate and confirm as much information as possible in writing. Set up checkpoints on progress, and keep notes on verbal agreements.
4. Personally verify that work is completed as intended, which includes the keeping of acceptable records. If possible, go over the completed work with the worker(s) before they leave the job site.
5. Assume that workers do not know the procedures in place to handle an emergency. The process operator and workers cover the safety rules on emergency procedures before work starts. The process operator as troubleshooter makes informal safety and housekeeping checks at key points from the start through the completion of the work.
6. Stay abreast of personnel changes. Know the personnel assigned to the area and their responsibilities. Make personal contact to ensure that shared responsibilities are understood.

3.6 TRUSTING INSTRUMENTS

Modern instrumentation can do a marvelous job of providing information needed to control a petrochemical process. Inclusive data from point sources can be measured and recorded. Control loops provide continuous adjustments to process conditions. The troubleshooter combines his knowledge of technology with the information available from modern instrumentation to detect early effects of a process upset in the making. He knows that any deviation from normal in an instrument readout has a cause that deserves an explanation. Error in an instrument reading can be checked against other readings, since there is considerable dependency among process variables. The wary troubleshooter trusts instrument readings that support his expectations based on experience and knowledge of process technology.

3.7 THE UNEXPECTED PROBLEM

Chemical plants are designed with controls to establish low-risk upper and lower limits on process variables, yet produce the desired physical and chemical changes. If a process variable strays outside on approved limit, there is a potential to damage equipment and create a hazard to personnel. The normal preventive measure is the installation of alarms on low and high control points set for process variables backed up by an automatic shutdown system if corrective action is not taken. The intent is to protect equipment and personnel from hazardous conditions that develop too rapidly for human intervention. This is an essential safeguard against a process change that occurs too rapidly for a process operator to detect, analyze, and take corrective action. However, the process operator as troubleshooter often prevents the need to energize alarms and automatic shutdowns through early detection and management of a process change discovered on plant patrol.

The probability that a sudden process change will escape detection and interrupt normal operation is a distinct possibility in a petrochemical plant. Fire, explosion, and a major release of a hazardous material cannot be ruled out as the result of equipment failure or human error. The process operator as troubleshooter is trained in emergency procedures to contain out-of-control conditions. His first responsibility is not to put himself or others at risk.

Determining the cause of a process upset may well depend upon the observations and records available on process conditions immediately before and during initial phases which only the process operator can provide. The effects witnessed between the onset of the unexpected problem and the initial action taken to contain it can provide valuable insight on the probable cause to the technical specialists on the troubleshooting team. Work begins on this very important troubleshooting effort as soon as process conditions are stabilized.

The shift log most likely provides the only credible and unedited documentation of effects that will ever exist on the initial conditions of a process upset. Hindsight will undoubtedly generate questions and requests for more information on the initial effects. The process operator as troubleshooter takes the responsibility to respond to all questions and to keep all information complete, correct, and current.

3.8 STEPS IN SOLVING THE UNEXPECTED PROBLEM

The unexpected problem occurs in three steps that the troubleshooting team reconstructs to define root cause. The first step is to identify by means of effects the interaction among people, places, and things. The second step includes analysis of the early effects before corrective action is taken. The third step is to reach a conclusion on root cause. All three steps are developed from information that answers the basic questions: What, when, where, and how did it happen?

In essence, finding the solution to an unexpected problem is reverse troubleshooting. The root cause(s) is determined by deduction from a plausible set of effects that created the upset condition. This leads to corrective action that eliminates the cause and reoccurrence of the problem after the fact.

3.9 SUMMARY

Detection and management of process problems make a good case for employing process operators who have a technical education. Broad technical knowledge is necessary to plan and execute a plant patrol aimed at the detection and communication of early changes in process conditions that precede a process upset.

There is a profound difference between the intrinsic and unexpected problem. The intrinsic problem is attacked with patience and close monitoring as progressive development occurs. Continuous but incremental improvement in management is seen as good troubleshooting. The unexpected change in process conditions has to be sized up quickly, and in many cases corrective action must be taken immediately since permitting further development increases the risk for equipment damage and potential hazard to personnel. After the fact, the unexpected problem is solved using effects to reconstruct the events to determine cause. The technical knowledge and communication skill of the process operator are put to the test as a sound source of information on the initial conditions of an unexpected problem.

CLASS EXERCISES

The three exercises in this chapter—Exercises 3.1, 3.2, and 3.3—are intended to demonstrate the detection and management skills needed among troubleshooters when working as a team.

The instructor divides the class into teams. The team organizes itself following the guidelines in Figure 1–1 to come up with the best mix of talent needed to attack the problem.

Both an oral presentation by each team and a written report by each team member are recommended as practice in communication skills. Grading takes into consideration readability, correct use of technical terms, spelling, and grammar.

Note: The instructor will be the technical advisor upon request.

■ EXERCISE 3.1 NEW REFRIGERATION UNIT STARTUP INSPECTION

Team Members **Team Leader**

________________________ ________________________

____

____

____

A contractor has completed the installation of a 200-ton/day refrigeration unit that works through the vapor-compression cycle. There are four basic steps shown in the figure below.

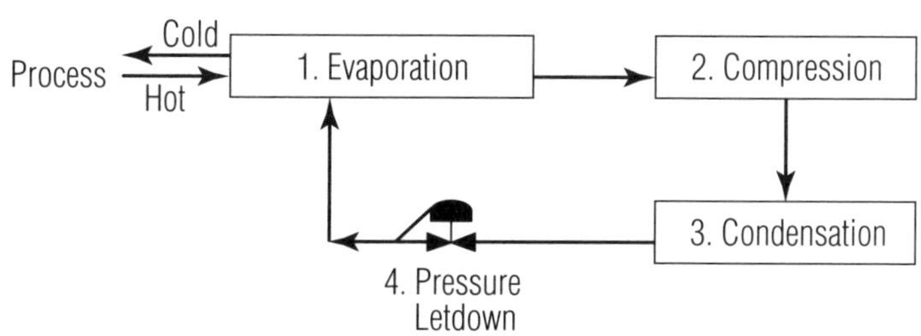

Use Figure 8–3–1 in Chapter 8 for more detail on the four steps in the vapor compression cycle. The exercise purpose is to choose a team and the steps in plant patrol that will assist the team in its inspection.

Instructions

1. The team is to make a startup inspection of a new refrigeration unit. Choose skills that would be appropriate to identify problems with piping, instrumentation, and equipment.
2. List the items studied under plant patrol that will be of most help in a successful inspection.
3. Each team is to explain its choice of personnel and the chosen plant-patrol items to the class. The class will discuss the choices.
4. Each team member submits a write-up to the instructor.

Time

Instruction 1	10 minutes
Instruction 2	10 minutes
Instruction 3	15 minutes
Instruction 4	Homework

■ EXERCISE 3.2 CONTRACT WORK

Team Members	Team Leader
____________________	____________________
____________________	
____________________	
____________________	

A crew of contract workers is scheduled to replace a tank in your area. The job will take four days and three nights.

Instructions

1. You are one of the four members of the relief team involved in around-the-clock coverage of the installation of a new tank. The purpose is to assist the contract workers by providing information and answers to their questions that will ensure a safe installation in the allotted time.
2. What patrol factors and suggestions under the human factor will be of primary importance in order to follow this job properly?
3. In an oral presentation, explain your thoughts to the class.
4. Write up your thoughts and submit to the instructor.

Time

Instruction 2	15 minutes
Instruction 3	15 minutes
Instruction 4	Homework

Note: The instructor is available to provide practical experience.

EXERCISE 3.3 CHRONIC SEAL LEAK

Team Members	Team Leader
____________________	____________________
____________________	
____________________	
____________________	

A centrifugal pump in your area has a shaft seal that fails frequently and a leak results. The leakage is toxic and flashes to vapor upon release of pressure as it enters the atmosphere. The seal is both a chronic maintenance problem and a safety hazard.

Instructions

1. Select team members to study the problem. Detection of a very small leak and proper corrective action are important objectives.
2. What items under plant patrol would you use to provide the operations team with the information needed to troubleshoot the problem?
3. What methods would you choose to communicate information on the problem?
4. Present the team's thoughts to the class in an oral presentation.
5. Write up your thoughts and present them to the instructor.

Time

Instruction 1	10 minutes
Instruction 2	10 minutes
Instruction 3	10 minutes
Instruction 4	15 minutes
Instruction 5	Homework

Note: The instructor is the process control engineer.

CHAPTER

4

Climates for Problem Development

LEARNING OBJECTIVES

- See prevention of problems as a troubleshooting responsibility.
- Use the functions of the operations team to troubleshoot a climate for problem development.
- Connect the steps in plant patrol to effective troubleshooting of a climate for problem development.
- Understand that when the climates of a problem are present, prevention becomes possible through troubleshooting.

KEY TERMS

Deadline
Workmanship
Root cause
Specification
Dummy run
Feedstock
Short cut
Installed spare
Performance parameter
Interaction
Resource
Checklist

Certain plant activities are prone to problems that repeat again and again without the intervention of a good troubleshooter. Included among this select group are changes in process technology, tight schedules with *deadlines*, operations that must work the first time, activities involving many people and many skills, communications at shift change, and responsibilities that are not specifically assigned or that can be misunderstood. In each climate, ample conditions are present that allow unexpected problems to develop, involving people, places, and things. The process operator as troubleshooter is aware of a climate for problem development in his unit and takes the troubleshooting action needed to prevent problems.

Deadline
A point in time when an objective is to be met.

Some potential problems are common to all plant activity. Operational errors, faulty communications, lack of teamwork, insufficient training, poor *workmanship*, and lack of knowledge of basic procedures can all contribute to problems the process operator as troubleshooter must recognize and prevent.

Workmanship
The quality of both mental and physical activity carried out in a professional manner.

Startup of new and untested equipment is a common source of problems. Equipment problems often have their *root cause* in the substandard performance of equipment after installation or maintenance. The necessary revisions and training in the standard operating procedures on new equipment can be overlooked or inadequate. The process operator as troubleshooter must counter with a plant patrol that closely monitors new equipment for proper installation and adequate performance *specifications*. He must have training in proper operating procedures.

Root Cause
The most probable effect behind a problem that, when eliminated, eliminates the problem.

Specification
A condition or property of a product that can be measured to ensure it meets preset limits.

Another factor that causes problems, unless carefully prepared for, is new technology. The process operator as troubleshooter needs to be educated in the new technology installed in his area before he has the online operating responsibility. He makes full use of available technical training on the chemical and physical changes taking place. A productive exercise is to use proposed operating procedures to *dummy run* new equipment on the blackboard, then do the same in the field with full participation by all the involved members of the operations team.

Dummy Run
Mentally working through a plant operation by simulation of the actual operation using non-toxic materials.

Another area that deserves attention by the process operator as troubleshooter is a change in process materials. Checking out the specifications on all imported materials before they are introduced into the process must be standard procedure. The process operator should have both vendor and his own laboratory checks on purchased materials in hand before approval for process use. Taking time to confirm specifications or to run confirmation checks on vendor data is good troubleshooting. A quality-control program on in-process materials and imported *feedstocks* that is comparable to checks on finished products is sound troubleshooting that uncovers potential problems for corrective action before they enter the process.

Feedstock
The prime material fed to a process.

Consistent discovery of problem effects is the reward of the process operator as troubleshooter, who expects the unexpected at all times. The troubleshooter who communicates his discovery of a change in a process condition immediately is a productive team player in preventing process upsets.

Troubleshooters who consistently review work reported as completed, monitor work in progress, and confirm preparations for future work detect many problems in daily activities that involve people, places, and things. The plant patrol is organized to troubleshoot climates for problem development and to consistently produce positive results in problem detection and management.

4.1 CHANGES IN TECHNOLOGY

No job, workplace, or product is immune from the progress of technology and the enhanced process performance that follows. Processes that the process operator as troubleshooter works with today will become obsolete tomorrow without the continuous influx of better technology. The process operator accepts the new technology incorporated into his unit as a good sign that his product will stay competitive in the marketplace.

New technology can be learned, ideally, in two steps:

1. Monitor the installation of new or modified facilities and seek technical explanations on the technology changes.
2. Write the standard operating procedures to establish standard operating conditions prior to startup.

As a troubleshooter monitors the installation of new equipment, the opportunity is there to think through possible startup and operational problems. He follows up with suggestions that can prevent those startup and operational problems and he volunteers to help implement his ideas.

The process operator as troubleshooter looks to the technical specialists to explain new and modified technology that is employed in new and modified processes. A discussion about new technology with a specialist provides an excellent opportunity for that process operator to extend his knowledge and to introduce his thoughts on possible startup problems. No question is too elemental to be asked and a sound, technical answer to be expected.

A discussion of process changes with process operators in other areas that could be impacted by new technology is solid troubleshooting practice. New technology can often trigger unexpected consequences in unexpected places, such as in power demand and waste-disposal facilities. Opening lines of communication on any process change prepares others for unexpected problems.

The process operator as troubleshooter takes the following actions to manage changes in technology with a positive effect on his job:

- Accept a change in technology as an opportunity to keep a process healthy and to extend its useful life. Expect revisions in current job responsibilities and volunteer to work on the details. A positive attitude promotes good teamwork.
- Participate fully in training and education programs. Offer to assume a leadership role. No student masters the subject better than the teacher does. Use "what if" discussions to expose possible startup problems and propose solutions. Concentrate on understanding the control logic and hazard prevention safeguards.
- Volunteer to follow installation of new equipment. Work on revised startup and shutdown procedures, safety rules, hazard identification, and other pre-startup activities. This approach provides excellent knowledge of new equipment and how it is expected to work. Troubleshooters who understand the basic features of equipment make good communicators as performance problems crop up.
- Work with technical and mechanical personnel on startup plans. This is an excellent opportunity to clarify control logic, alarm and interlock settings, servicing of equipment, and the recording and logging of data. It brings the process operator on board as a technical team player on preventing equipment problems.
- The process operator of new or modified facilities looks forward to establishing his role as the foremost expert in the operation of the equipment. It follows that he must learn the technology behind the equipment to become a respected member of the troubleshooting team.

4.2 TIGHT SCHEDULES WITH DEADLINES

Many plant activities must be carried out on a tight schedule with a deadline to minimize downtime. This implies an operations team in a hurry, which is a climate for problem development.

A tight schedule with a deadline presents all the problems associated with hurried decisions, garbled communications, *short cuts*, and make-do workmanship. Meeting a tight schedule starts with a well-planned shutdown and ends with a well-executed startup. Those process-operator responsibilities are easily met through teamwork by the relief team. Activity in between shutdown and startup is a climate for things to go wrong.

Short Cut
The act of bypassing established procedures with the intent to reach an end result quickly.

The process operator as troubleshooter expects that a tight schedule with a deadline will be a climate for problem development from start to finish. He constantly expects the unexpected. He works with the operations team to uncover effects of potential problems as work proceeds to eliminate the causes. Major jobs are carefully planned during sessions with all the parties involved. A process operator and a mechanic doing a walkthrough can be an effective problem-prevention tactic on small jobs.

The process operator as troubleshooter can contribute to the management of tight schedules with deadlines in several key ways:

- Obtain a copy and study the proposed schedule of all activities. Carefully plan to meet your responsibilities for a timely shutdown and efficient startup. Look for ways to assist others in their work. Comparing notes with those scheduled to do the work will invariably expose conflicts that can be prevented. That is fine troubleshooting.
- Follow standard startup and shutdown procedures. Short cuts or impulsive changes in standard practices to save time are serious problems that lead to unforeseen causes of additional problems.
- Continuously monitor all activity in the area. Ensure that the steps in the schedule that are your responsibility stay on track.
- Share any change from what is set down in the schedule with others on the operations team immediately. Corrective action more than likely will take teamwork.
- Start your relief off with a complete status report on work completed, work in progress, and the preparations needed to keep future work on schedule. Set priorities on the actions your relief must take for several hours into his shift.

4.3 OPERATIONS THAT MUST WORK THE FIRST TIME

There are routine plant activities—such as switching equipment in and out of service, changing to a new source of feedstock, and coordinating action with other process areas—that must work the first time or result in a process upset. The troubleshooter prevents a number of unexpected problems in this climate for problem development by following standard operating procedures. Short cuts are attractive, but they are difficult to communicate effectively from shift to shift and are indefensible when they go wrong.

The troubleshooter routinely checks the condition of *installed spares* for operability. An inoperable spare on standby is an unacceptable compromise of the backup insurance that was considered necessary to reduce losses in the event of equipment failure.

Installed Spare
An exact duplicate of an equipment piece that permits a change in service if needed.

Performance Parameter
A defined requirement in the output of an equipment piece.

The *performance parameters* of replacement equipment are checked out by the prudent process operator as troubleshooter prior to installation, and operability is tested by a short run prior to a need to be placed in continuous service. This is a good opportunity to team up with a technical specialist and confirm that equipment meets the specifications to perform at standard operating conditions.

A review of data sheets on catalyst, acid, salt, and other solutions to ensure proper concentrations is mandatory before the material is introduced into the process.

Any change in a feedstock brought in from outside or a switch to a previously stored in-process material entails some risk for an unexpected process upset unless the material meets specifications. The process operator as troubleshooter checks the specifications on material prior to introduction into the process. If necessary, he may request assistance from the control laboratory. A process upset in one area can produce off-quality feed for units downstream. Any change in an in-process stream outside specifications must be communicated so corrective action can be taken.

A third common cause of problems in an operation that must work the first time is reliance upon another person or a function carried out in another area. Often a coordinated effort is necessary. The troubleshooter accepts that effective (even redundant) communications are called for to ensure a properly coordinated operation that involves two or more persons or a remote area. One sure need is adequate lead time to set things up.

The process operator as troubleshooter can detect and manage problems resulting from operations that must work the first time by use of the following steps:

- Follow the installation and maintenance of replacement equipment carefully. Compare performance specifications with the equipment being replaced.
- Identify any inoperable spare equipment, and enter the repairs needed into the shift log. Start the paperwork for repairs if necessary, and follow up to ensure a prompt response.
- Confirm that any scheduled change in a feedstock is backed by documented analysis against specifications. Maintain a complete file as a reference. Alert operations downstream of any change in the composition of material produced in your area.
- Communicate frequently on any activity shared with others. Provide ample lead time when requesting assistance and follow up on results. Stay in touch with all the parties involved until the job is completed. If you are the party downstream, check to ensure the timing has not changed.

4.4 ACTIVITIES INVOLVING MANY PEOPLE AND MANY SKILLS

Interactions among production, mechanical, and technical teams must be carefully planned in the operation of a petrochemical plant. Collectively, these teams are made up of specialists in operations, safety, environmental protection, equipment, quality control, instrumentation, and *resource* management who can all be involved in a major activity such as the annual turnaround. The process operator fills a logical need as the operations expert who is in the position to detect, analyze, determine cause, and communicate the corrective action necessary to prevent problems among activities that involve many people and many skills. The corrective action often is simply to bring the problem to the attention of the right troubleshooters responsible for schedules and timing.

Interaction
Two or more activities having an effect on each other that must be coordinated.

Resource
Manpower, materials, and facilities available to carry out a specific task or mission.

Another practice that leads to problems involves people provided through contract with outside service organizations. Expert services on selected processes, such as refrigeration and water treatment, are commonly provided under contract. Heavy maintenance is often contracted to outside specialists to better utilize manpower and to provide the heavy equipment needed. These workers, who are skilled specialists, appreciate the process operator who contributes his knowledge of the process, equipment, work practices, and hazards to help them perform their work effectively and safely.

Activity among many people and many functions peaks during the annual turnaround and is also a characteristic of startup of major new facilities. The plant is literally covered with workers selected from the operations team and others on contract during these periods of intense activity—a true climate for problem development.

Activities that involve many people and many skills involve a number of personnel from the operations team such as electrical, instrument, and environmental specialists, who have plant-wide responsibility to provide coverage on an as-needed basis. The process operator as troubleshooter recognizes the fact that these people do not know the process. He calls upon his knowledge of the process to assist them in their work.

The process operator must not underestimate the importance of his ability to provide input on safe working conditions and practices in his area for all workers. Expert workers from outside are not expert on the hazards in every area they work in, so a knowledgeable process operator is a reassuring presence. He relays to all workers the emergency procedures and escape routes that are specific to the area. He must conduct himself as the resident expert.

The process operator as troubleshooter fills the role of observer of the activities among all workers and helps them to follow acceptable safety and work practices. The process operator serves as communicator to keep all groups aware of the activity of others and the hazards in place at all time. The classic example is to prevent working under the path of a crane.

The responsibility of the process operator to cover activities involving many people and many skills can be met through several specific actions:

- Review the schedule of work in the area before the work begins. Confirm what the workers intend to do and when they intend to do it. Follow the work in progress and stay available to answer questions. Look for conflicts with other planned activity in the area.
- Insist on following accepted work practices and safety procedures. There is rarely an acceptable exception. Take the initiative to review known hazards with all workers. Teamwork prior to start of work prevents needless misunderstandings.
- No change is to be made in the scheduled work without prior approval. A change that would delay completion of a planned activity can be a significant problem. Thoroughly inform your relief of all activity that needs his attention—even to the point of providing a *checklist*.
- Keep all members of the operations team with a need to know informed on the status of all activity. Information of a change or delay in scheduled work is communicated immediately to the appropriate troubleshooting team for resolution.

Checklist
A list used as a reminder of tasks one is responsible for completing.

4.5 RESPONSIBILITIES NOT SPECIFICALLY ASSIGNED OR MISUNDERSTOOD

The importance of good communications cannot be overstated. A common difficulty in good communications is to confirm that each assignment and the responsibility to carry it out are fully understood and accepted by every worker. A clearly understood commitment to complete an assigned task by each member is essential to a productive organization.

Troubleshooting is, above all, a team effort that relies upon communications among the mix of technical expertise and operational skills that are needed to attack a problem. The process operator sets the stage by thoroughly understanding and accepting his responsibility to communicate on the activity in his area with those who may be affected.

It is often difficult to draw clear lines of responsibility since many shared activities overlap or are carried out in stages. Schedules are built around the need to work as a team in the proper sequence. Communicating a clear schedule on shared activities is the key to efficient use of resources like manpower and equipment.

A workable technique to clarify individual responsibility is to describe how one intends to approach his part of the job. If there is conflict with what others plan to do, they will speak up, and a conflict is prevented. This is a useful tactic on any activity to help clarify job functions before a job is started and eliminate confusion later. Others are likely to make a commitment if you make your own intentions clear.

Here are several suggestions that will help eliminate confusion concerning responsibilities not specifically assigned or misunderstood:

- Make personal contact with every worker with whom you share a responsibility an absolute must before work starts. Clarify what each of you is committed to do. Agree that you will communicate through checkpoints on progress.
- If person-to-person contact is impossible, make an extra effort to communicate your commitment and your expectations of a commitment from others in the shift log. Fully communicate to your relief what he is committed to and why. Be specific on who he is scheduled to work with.
- Contact others involved immediately on any change on your part in the status of a shared activity.

4.6 COMMUNICATIONS AT SHIFT CHANGE

To ensure continuity of operations, consistent thought and effort are needed to communicate the information and priorities on activities during transfer of responsibility at shift change. During that few minutes of verbal exchange, the process operator must summarize his activity and instruct his relief on the activity that needs his attention.

A checklist helps to organize activity in a logical timetable (Figure 4–1). The checklist itemizes the activity, location, and responsible group/person in a time frame.

A large amount of information can be summarized, amended, added to, or corrected by use of an easy-to-prepare checklist. The checklist is an excellent tool to cross check the status of a large number of related activities by intercom or radio. Any shortfall in progress on a job is easily noted with

I Activity	II Location	III Responsible group and contact	IV Time start and finish	V Comment

Figure 4–1
Checklist Form

the name of a contact to facilitate follow-up on corrective action. A checklist is also a valuable source of recall as the work progresses.

In order to communicate effectively at shift change, use the shift log for detailed information and use verbal exchange to give short-term instructions and to set priorities. This approach is based on the premise that a large quantity of detailed information cannot be exchanged and retained in a few minutes of conversation, but the need to set priorities and assign short-term action can be.

Shift-change communications should conform to the following guidelines in the shift log and during the verbal exchange:

- Communicate information in the shift log, where accuracy and detail are important. Numerical data are a prime example.
- Document in the shift log the progress on activities and commitments with deadlines. Enter name and functions of involved team members. Close out activities that are completed.
- Introduce in the shift log new activities that require follow-up. Any changes or amendments to plans, data, or schedules from previous logs should be highlighted.
- Communicate verbally the priorities requiring immediate action and follow-up by your relief during the first two hours of the shift.
- Communicate verbally the need to monitor activity with stop-go check points.
- Call attention verbally to reports, memos, and other sources of information that should be read and understood.
- Highlight verbally any potential safety concerns or hazardous conditions that need attention.

■ 4.7 SUMMARY

Climates for problem development bring out the point that one operates a petrochemical plant in a dynamic environment of machines and people. There are many teams available to keep the operation running smoothly and safely, which requires each team member to fully understand his role and accept his responsibility to carry it out.

The process operator is not a silent bystander as others work in his area. He knows the schedule, he knows the workers, he knows the process, and he recognizes the climates that bring on problems if not detected and managed properly.

CLASS EXERCISES

The five exercises in this chapter—Exercises 4.1, 4.2, 4.3, 4.4, and 4.5—are intended to demonstrate the detection and management skills needed among troubleshooters when working as a team.

The instructor divides the class into teams and assigns one as a team leader. The purpose of the exercises is for each team member to choose an action from Table 4–1 on the climate for problem development in the assigned exercise. The team consolidates each member's actions into a composite of team actions.

1. Each student follows the specific instructions in the assigned exercise.
2. The team leader pulls together the input from each team member and prepares a summary for presentation to the class.

Note: The instructor will be the technical advisor.

Table 4–1
Troubleshooting Activities

1. Read shift logs.
2. Keep the shift team informed.
3. Check work in progress.
4. Check work completed.
5. Check future work sites.
6. Review lab results and memos.
7. Select a specific area to inspect.
8. Learn the mechanical contacts in your area.
9. Learn the technical contacts in your area.
10. Expect the unexpected in all activity.
11. Complete safety and housekeeping inspections.
12. Study work schedule and work orders.
13. Monitor and report on area activity.
14. Take notes throughout the shift.
15. Question strangers on their knowledge of area safety rules.
16. Set priorities at shift changes.
17. Schedule a walkthrough with team members.
18. Check equipment specifications.
19. Compare what is to what should be.
20. Sketch your own P&I diagram as you trace out the piping in the field.
21. Check out the communications equipment.
22. Plan an effective plant patrol.

EXERCISE 4.1 NEW TECHNOLOGY

Team Members **Team Leader**

______________________________ ______________________________

____

____

____

You have been transferred to a new job running a liquid-liquid extraction unit. You look forward to learning the technology at work in your new assignment. You will train with an experienced process operator for three weeks before you work the first shift on your own. You know that the three weeks of training will prepare you to operate the process as long as you follow standard procedures and ask questions when in doubt. Your objective is to be a good troubleshooter as well.

Instructions

1. Choose *four* actions from Table 4–1 that you will concentrate on during the three-week training period to assist you as a troubleshooter.
2. Choose *four* actions from Table 4–1 that you will concentrate on after you take over the unit on your own to be a troubleshooter.
3. Work with the team leader to prepare an oral presentation to the class.
4. Write up items 1 and 2 for the instructor.
5. The team leader writes a summary of the oral presentation for the instructor.

Time

Instruction 1	10 minutes
Instruction 2	10 minutes
Instruction 3	10 minutes
Instruction 4	Homework
Instruction 5	Homework

EXERCISE 4.2 TIGHT SCHEDULE WITH DEADLINES

Team Members	**Team Leader**
____________________	____________________
____________________	
____________________	
____________________	

It is 5:00 A.M. A contract crew is scheduled to replace two fan motors and to complete a list of general-maintenance items on the cooling tower in your area at 8:00 A.M. You are running an hour behind on the scheduled shutdown of the tower. Your relief is due at 6:00 A.M. and must make up the time. How can you help him out?

Instructions

1. Which four steps from Table 4–1 would you select to get your relief off to a good start? Explain your reasons. Would you add an item to the list?
2. In an oral presentation to the class, present your choice of actions and your reasoning.
3. Write up your work for the instructor.
4. The team leader will write up the consensus of the team.

Time

Instruction 1	10 minutes
Instruction 2	5 minutes per team member
Instruction 3	Homework
Instruction 4	Homework

EXERCISE 4.3 OPERATIONS THAT MUST WORK THE FIRST TIME

Team Members **Team Leader**

____________________ ____________________

____

____

____

A pump has been delivered to your area for installation tomorrow.

Instructions

1. What four actions would you take from Table 4–1 to ensure that the new pump will perform as needed online?
2. Present your choice of action and reasons to the class.
3. The team leader presents the consensus of the team to the class.
4. Write up your presentation for the instructor.

Time

Instruction 1	10 minutes
Instruction 2	5 minutes per team member
Instruction 3	Homework
Instruction 4	Homework

EXERCISE 4.4 ACTIVITES INVOLVING MANY PEOPLE AND MANY SKILLS

Team Members **Team Leader**

____________________ ____________________

____

____

____

You are in the third day of a ten-day shutdown and assigned to a special day shift with hours from 6:00 A.M. to 6:00 P.M. Most of the work is scheduled on your shift by workers that start at 8:00 A.M. and knock off at 4:00 P.M.

Instructions

1. What four actions from Table 4–1 could you take on between 6:00 A.M. to 8:00 A.M.?
2. What four actions from Table 4–1 would be on your checklist at 8:00 A.M.?
3. Present your checklist to the class with your reasoning.
4. Write up your presentation to the class for the instructor.
5. The team leader will present the consensus of the team.

Time

Instruction 1	10 minutes
Instruction 2	10 minutes
Instruction 3	5 minutes per team member
Instruction 4	Homework
Instruction 5	Homework

■ EXERCISE 4.5 COMMUNICATIONS AT SHIFT CHANGE

Team Members	Team Leader
______________________	______________________
______________________	
______________________	
______________________	

Below are notes that you collected during your shift.

Shift Notes

6:00 A.M. to 6:00 P.M.

6:30 A.M. Discussed compressor job with mechanic. Need parts. Delay until next shift.
7:00 A.M. No. 1 reactor feed pump has signs of a small seal leak. Will follow up.
8:45 A.M. Took set of data on topper column as requested by process control engineer. Wants another set next shift.

Feed	30,500 pph
Tails	5,250 pph
Make	25,100 pph
Base temp.	210°C
Head temp.	180°C

9:00 A.M. Water flow on refrigeration condenser 200 gpm. Scheduled acid clean for Mon. night.
10:00 A.M. Supv. complained about housekeeping around pelletizer.
3:00 P.M. Checked out the reactor pump seal with a mechanic. No change.
4:00 P.M. A pre-planner dropped off the turnaround schedule for comments. Will you work on it tonight?
5:00 P.M. The CCR operator thinks we will make a product change around 3:00 A.M. The supervisor will confirm.

Instructions

1. Identify the information you would write in the shift log.
2. What information would you communicate verbally?
3. Explain your choices to the class.
4. The team leader will present the consensus of the team to the class.
5. Write up the presentation for the instructor.

Time

Instruction 1	10 minutes
Instruction 2	10 minutes
Instruction 3	5 minutes per team member
Instruction 4	10 minutes
Instruction 5	Homework

CHAPTER

5

Problem Solving as a Process

LEARNING OBJECTIVES

- Understand the connection between troubleshooting and problem solving.
- Recognize the need for teamwork in problem solving.
- Understand that teamwork is required for the problem-solving process to be effective.
- Describe the differences between deductive and inductive reasoning.
- Explain the safety problems involved in trial and error as a means to solve problems.

KEY TERMS

Solution
Corrective action
Effect
Result
Probable cause
Root cause

Solution
The action taken to eliminate the cause of a problem.

The process operator, with an education in process technology, is in the position to make a significant contribution to the *solution* of process problems. As a troubleshooter he initiates the timely information on process conditions that feeds the problem-solving process. Connecting problem solving to troubleshooting and how the problem-solving process is applied is the first objective of this chapter (Table 5–1).

Traditionally, technical experts take the information on process conditions provided by the process operator and go about their business of solving problems. Many are problems that would be well within the capability of a technically educated and well-trained shift team to solve and to implement the solution. The technically educated process operator and the technical experts can team up on the solution to complicated problems.

Troubleshooting and problem solving have features in common, but each is less effective without the support of the other. Troubleshooting is

Table 5–1 Comparison of Problem Solving and Troubleshooting

ACTIVITY	PROBLEM SOLVING	TROUBLESHOOTING
The Unexpected Problem		
1. Patrol for early detection of abnormal process conditions (effects).		X
2. Communicate with the shift team, and request assistance if needed.		X
3. Take corrective action.	X	X
4. Document effects and action taken.	X	X
5. Follow up and report results of corrective action to the operations team (shift log).	X	X
6. The operations team analyzes the effects, the corrective action taken, and the results to ensure that root cause is eliminated.	X	
7. If needed, further study of the problem is recommended.	X	
The Intrinsic Problem		
1. Monitor and report the progressive effects.		X
2. Predict the need for and initiate corrective action that minimizes loss of production.	X	X
3. Report any new information on intrinsic problems such as a change in development, physical appearance, location, etc.		X
4. Implement actions approved by the operations team to alleviate the progressive effects of the problem as they are deduced from new information on *effects*.	X	
5. Follow up and document *results* on action taken to alleviate effects, which in itself is new information.	X	X

Effect
A measured, observed, or sensed abnormality in the interaction of people, places, and/or things.

Result
The change produced by a deliberate action.

justified on the basis that early detection of an abnormal process condition creates the opportunity for *corrective action* that prevents a process upset. Problem solving is justified on the basis that unless the root cause of an abnormal process condition is exposed and eliminated, the problem will reoccur.

Corrective Action
A change that is made to minimize the effects of a process upset and that is aimed at eliminating root cause.

5.1 INDUCTIVE AND DEDUCTIVE REASONING

Training is how the process operator learns to use his knowledge and skills to complete specific tasks. Training and experience combine to analyze what is compared to what should be in process conditions. Experience often fine-tunes knowledge and skills into a proven combination that allows one to explain the reason behind a change he has encountered before. Through experience one comes up with the corrective action needed to bring an errant process condition back into control. Behind the process change selected to control is the mental process called inductive reasoning, and it works well to manage repetitive problems.

What happens when a situation comes up that is outside one's training and experience or the effects have a cause not seen before? The operator faces a set of conditions (effects) for which the cause is not obvious. The first question that comes up is "What could have caused this?" Would it be reasonable to say that several *probable causes* would come to mind? Would it be reasonable to compare the effect to the list of probable causes and find one that explains more of the effects than any of the others. Once the most probable cause has been identified, would it be reasonable to plan and take corrective action to eliminate it? The mental process one goes through to plan and take corrective action on a situation outside one's experience and training is called deductive reasoning.

Probable Cause
The most likely effect that initiated the problem.

The troubleshooter strives for the optimum practice of using both inductive reasoning (training and experience) and deductive reasoning (problem-solving process) when solving both unexpected and intrinsic process problems.

5.2 TRIAL-AND-ERROR SOLUTION TO A PROBLEM

One problem-solving process that is resorted to in the absence of an understanding of the process technology at work is trial and error (Table 5–2). Trial and error is easy to implement, and produces a fast result—good or bad. The drawback is that there is often insufficient information and knowledge at work to assess all the consequences. This is a potential compromise

Table 5–2
Trial-and-Error Solution to Problems

1. Make a change.
2. Observe the results. Decide: good or bad.
3. Make another change to eliminate a bad result.
4. Observe the results. Decide: good or bad.
5. Repeat steps 3 and 4 until the overall results are acceptable.

of well-established safety and operating procedures and is totally unacceptable in an industrial workplace. The elements of guesswork and use of assumptions found in trial and error are more useful to exploratory work than to a commercial operation.

Trial and error is attractive in tight situations. The process operator as troubleshooter is expected to take quick action on unexpected changes in process conditions to prevent a process upset. This is where deductive reasoning must prevail. Time must be taken to match the effects to a probable cause and to think through the consequences of corrective actions. If not, the action taken will be open to creating new conditions that are even more difficult to correct and this is not acceptable.

5.3 SCIENTIFIC METHOD IN THE SOLUTION TO A PROBLEM

The use of the scientific method to explore the unknown is much less risky than trial and error and is very effective as a laboratory tool (Table 5–3). Experiments are based on partly known technology, and the results are anticipated. Adaptation of the scientific method to a commercial operation is difficult because the equipment is large and reacts to change slowly. Also, the petrochemical plant is not a cost-effective environment to deal with technology that is in some experimental stage of development. Running multiple experiments that may give mixed results (no matter how well thought out) is difficult to support financially in a commercial operation.

This does not rule out the scientific method entirely to improve petrochemical plant performance. An experimental operation that offers large rewards and low risk may be acceptable in a plant test.

Table 5–3
Scientific Method in the Solution to Problems

1. Design and carry out an experiment to produce a specific result.
2. Observe and analyze the result. Fully use technology as the basis.
3. Deduce the most probable cause for a less-than-expected result.
4. Redesign the experiment using knowledge gained from the first effort to produce an expected result.
5. Repeat steps 2, 3, and 4 until the desired result is obtained.

5.4 PROBLEM-SOLVING PROCESS

Until the *root cause* is found and eliminated, one can assume that a process upset will repeat. That is motivation for the operations team to have the technical resources in hand to troubleshoot abnormal process conditions. The problem-solving process that is offered here is an analysis of effects to deduce the probable causes of an undesirable interaction among people, places, and things (Table 5–4). Analysis is dependent upon extensive teamwork within the operations team, which processes the raw data provided primarily by the process operator.

Root Cause
The most probable effect behind a problem that when eliminated, eliminates the problem.

A thorough analysis of the effects to expose the cause through use of a unified problem-solving effort requires the input of selected technical expertise. Complete mastery of the ranges of process technology in use in a modern petrochemical plant is beyond the knowledge provided in any one technical field.

Petrochemical processes that require interpretation of technology by a number of specialists in order to solve problems make working in teams a necessity. A workable problem-solving process is necessary to hold the

Table 5–4
The Problem-Solving Process

The unexpected problem has been contained and is under control, but the problem is that the root cause is not known. Solving intrinsic problems takes a continuous effort. The problem-solving process is designed for both the unexpected and the intrinsic problem. Troubleshooters work as a team through the following steps:

DOCUMENTATION OF EFFECTS

What has happened, and on what object did it happen?
Where did it happen, and where is the object it happened to?
When did it happen, and when did it start?
How did it happen, and how big is it?

EVALUATION OF DATA

Seek out gaps and contradictions in the documentation of effects.
Identify distinct and unusual effects.
Reject guesswork, conjecture, and unsupported assumptions.

ANALYSIS OF THE EFFECTS

Compare what is to what should be.
List probable causes through the use of deductive reasoning.
Choose the most probable cause through the best match with the effects.
Prepare a list of solutions with alternatives.

CHOOSE A SOLUTION

The solution must be doable and technically sound.
The solution must eliminate the problem.

team effort together. The process operator is the specialist in detection and communication of process conditions. He needs to understand how his contribution fits into the problem-solving process.

Documentation of Effects

All problem-solving teams, no matter how well-staffed and motivated, fully understand that success comes from the quality of the information available to work with. The process operator as troubleshooter accepts his responsibility for the quality of the information on process conditions that only he is in a position to provide. An excellent start on communicating complete data is achieved by the process operator through the answers to four basic questions:

- *What* happened, and on what object did it happen?
- *Where* did it happen, and where is the object it happened on?
- *When* did the problem start, and when was it detected?
- *How* did the problem happen, and how big is it?

The major contribution by the process operator to problem solving is related to his effectiveness as a communicator in answering the questions above. Understanding several basic factors will contribute to his skill in achieving thorough documention of effects:

1. Only the troubleshooter who detects or discovers the problem can report accurately on the original effects. Later reports, no matter how convincing the inductive reasoning may be, are tempered by the original action taken to contain the problem.
2. No detail is too small to report, including all the actions taken to contain and control a problem. Notes made during the containment and control phases will make an important contribution to total recall in a timetable.
3. Do not hesitate to correct and amend the original information. Data reported in good faith can be seen in new light as new effects and additional evidence surface. However, do not change original data without a sound reason.
4. Speculation about the causes of the problem by the process operator as troubleshooter is acceptable. A plausible use of effects that is based on process technology to propose cause is especially well-received.
5. Specify causes and effects that were modified or eliminated by the actions taken to contain and control the problem.
6. Date all data and identify the source.

Evaluation of Data

The problem-solving team starts its work with a confirmation that the effects in hand are fully documented in sound technical terms and units. Descriptive statements are quantified as much as possible regarding time and place. The problem-solving team ensures that all reports on the problem and the action taken by all the troubleshooters of record are in hand. This assurance that the best and most complete story of what, when, where, and how this event happened is a critical first step in solving any problem.

The problem-solving team organizes the effects into a time frame that creates a logical reproduction of the changes in process conditions starting with the earliest known effect. Identify as guesswork, conjecture, and assumptions all data not verifiable.

When all the information is in hand, and before work begins on producing a list of probable causes, a helpful step is to precisely describe the problem in writing. This serves to focus the team effort on the cause supported by effects related to a clearly defined problem. To be of most use to the problem-solving team, the description of the problem is simple and specific. A good rule is that each independent change from normal process conditions should start out as a separate problem. This approach brings out independent problems hidden in a large number of reported effects. It also exposes the effects of several problems in series—the domino effect. Complicated technical problems become solvable only when broken down into logical pieces, which can be fitted together later.

It may be necessary to revise, amend, or clarify the description of a problem as additional information becomes available. Again, the objective is to keep each problem description connected to one set of effects. Two standards concerning the content of the written description of the problem are essential to produce the desired specificity:

1. The description of the problem covers one deviation from normal in specific terms that answer the four basic questions: *What*? *Where*? *When*? *How*?
2. The description of the problem deals with one set of related effects.

Analysis of the Effects

Deductive reasoning affords the objectivity needed to compare effects with probable causes and choose a root cause with a high level of confidence. Analysis of the effects must support a cause that is supported by a sound technical explanation.

Analysis is the time for creativity and imagination. The questions of *what*, *where*, *when*, and *how* are addressed in depth and from all angles to produce a comprehensive list of probable causes. The problem solvers must not judge the merit of each probable cause at this point. All probable causes that relate to one or more of the effects are treated as potentially valid. Avoid the overuse of inductive reasoning based on training and experience as a means to quickly disqualify the less obvious causes. Do not abandon the problem-solving process because one probable cause is immediately supported by a reasonable technical explanation. This is precisely the situation where a new insight into the problem is needed, or perhaps where more than one problem is involved.

When effects clearly support several probable causes, this is clear evidence that more than one problem is involved. If so, a written description of separate problems should be considered and the first problem description confined to keep the problem-solving team focused. Problems allowed to become too broad or too vague most likely cannot be solved because there is no agreement possible on one, single, definable root cause that can be eliminated. Breaking complicated problems down into manageable pieces is good problem-solving practice.

Deductive reasoning is the key ingredient in the analysis used to organize the connection between an effect and a cause. Deductive reasoning is used to make a rigorous analysis of what is to what should be. All causes must be supported by the technology that would create the effects. Inductive reasoning based on experience must also be supported by the technical facts. Be open to the possibility that a common process effect has several probable causes.

Deductive reasoning works best in an analysis of effects in search of causes that are fitted into a time frame. The early causes and effects often expand into new effects and causes of the problem at a later stage of development. Effects that identify the root cause of the problem are easier to coordinate if they are connected in time. The root cause that initiates a problem most likely starts a sequence of causes and effects that must be traced back to time zero.

Choose a Solution

The analysis of the effects has resulted in a list of probable causes. Hopefully, most of the effects can be matched to a specific or root cause. Remember that the root cause must provide a technically supportable answer to these questions: what, where, when, and how did the problem come to be? If the answer to those basic questions stands out among a list of probable causes, the root cause most likely has been exposed.

Implement the Solution and Write the Closing Report

There may be several alternatives to the first choice of action taken to eliminate the cause and problem. The following conditions are recommended before implementing the corrective action:

1. The corrective action must be doable in the opinion of the problem-solving team.

2. There is consensus among the team members that the resources needed to implement and sustain the corrective action are in hand or can be specified.

A report on the effort to solve a process problem in a complex operation is a valuable record that will prove useful to prevent duplication of work by future teams. A well-written closing report is also worked into indoctrination and training programs for inexperienced troubleshooters.

(*Note:* All troubleshooters should read closing reports, incident investigations, and technical reports on previous efforts to solve intrinsic problems. The objective is to utilize the problem-solving efforts of the past to prevent wasteful duplication of effort.)

5.5 SUMMARY

The intent of any process to solve problems must fill several specific goals. First, all the problem solvers are unified through use of the problem-solving process. They see the logic of working together to combine their expertise, which will help them as troubleshooters and problem solvers. Secondly, the problem-solving process promotes objectivity, which deters foregone conclusions of the solution based on experience alone.

In addition, the problem-solving process is useful to the process operator as he considers the effects of an unexpected problem while having limited time to decide on the probable cause before taking corrective action.

5.6 EXERCISES

The exercises in Chapter 8 are designed to support the learning objectives of Chapters 5 and 6. The exercise chosen is best completed in steps that match the steps in the problem-solving process as they are presented in class.

CHAPTER

6

Troubleshooting in the Team Environment

LEARNING OBJECTIVES

- Combine the lessons learned in Chapters 1 through 5 into an understanding of the process-operator role as a troubleshooter in the team environment.
- Know your fellow troubleshooters and their fields of expertise, which facilitates obtaining technical assistance.
- Understand that communication connects troubleshooters together. Through sharing of mutual information, troubleshooting succeeds as a team effort.
- Know that the process operator is the foremost detector and manager of abnormal process conditions, which clarifies troubleshooting responsibility.
- Know that the vigilant troubleshooter recognizes the climates where problems are likely to develop, which initiates a higher rate of detection and management.
- Know that problem solvers and troubleshooters perform better by working in teams that follow a logical problem-solving process, which best utilizes all the available expertise to solve the problem.
- Know that working in teams is essential to consistently troubleshoot and solve problems in a sophisticated workplace.

Modern chemical processes are technically sophisticated entities comprised of equipment, piping, instruments, and the personnel to operate, maintain, and engineer them. Multidisciplined, troubleshooting teams are a proven business tactic needed to detect early effects and prevent process upsets in this complex workplace. Becoming a skillful, well-trained process operator as part of a high-tech operations team is a qualification for the job. Also, making a competent process operator is a good start to becoming a troubleshooter. Add an education in the process technology in support of the training and skills of a process operator, and the process operator as troubleshooter is on the job.

This chapter summarizes troubleshooting principles into a cohesive pattern that the process operator can apply to troubleshooting in a team environment.

6.1 BECOME A SKILLED, WELL-TRAINED PROCESS OPERATOR

As a member of a shift team, the process operator consistently uses standard operating procedures to establish standard operating conditions—and rightfully so. Universally accepted procedures are needed to run plants through organized operation teams that must work together in harmony.

The process operator who is educated in the process technology has the mental resource to fill the first need in troubleshooting, which is to supply a technical interpretation of process conditions. Troubleshooting in teams relies upon a source of accurate data on process conditions that only the process operator is in the position to provide. Even the smallest deviation from a normal condition may be the initial effect that leads to an unexpected process upset. The process operator is continuously comparing what is to what should be with all the activity in his area.

The process operator in the team environment can expect to be asked for additional detail on the process information that he provides. Probing questions often expose an overlooked piece of information that is important to a technical specialist. Probing for details on a problem confirms the inclusiveness of all the available data and provides common agreement on its meaning. This follow-up to compare information provided by several troubleshooters encourages debate and critical thinking.

The good process operator is a team player who understands that an abnormal condition in his area can upset operations upstream and downstream. He must consider the impact of the conditions and the impact his actions to correct them will have on others. Specifically, the process operator avoids unilateral change. He communicates an abnormal condition and the corrective action being considered to the entire shift team as quickly as possible.

This ensures that any side effect created by an action to correct a problem is anticipated, which reduces the chance of a process upset elsewhere. The CCR operator can provide valuable assistance as a communicator among shift team members involved in a mutual activity.

There are operations outside the process areas that supply essential services such as steam, water, power, and raw materials. Planned changes in a process that could have an effect on an outside operation that provides a service should be communicated with a generous lead time so that preparations for the change can be made.

The shift team is, of necessity, a highly trained team of five to eight process operators. As a team, they develop working relationships based on mutual trust that reduces the possibility for human error. However, the probability is good that at least one shift-team member will be off the job on vacation, on special assignment, or indisposed. In most plants, members of shift teams are cross-trained with the capability to fill in on several jobs.

This provides the needed flexibility to fill a vacancy with a relief operator who may be rusty.

An operator brought in to cover a job for a missing regular is a classic climate for problem development. Good troubleshooting requires an extra effort by the entire shift team to fully support the new member, which reduces the chances for human error.

The process operator coming to work must be thoroughly briefed by the process operator leaving the job. The relief operator expects a well-prepared exchange of information that will get him off to a good start in managing current activities and alert him to future activity that requires his attention. A proven technique to meet this objective is for the operator leaving the job to set priorities on activity in the area for a few hours into the shift. This provides what the oncoming operator needs most: guidelines for his short-term responsibilities that sustain work in progress without interruption. This provides time for the relief operator to take stock and to prepare, on his terms, for activity scheduled later in the shift. This ten-to-fifteen-minute transfer of responsibility between process operators is an important opportunity to troubleshooting as a team.

The process operator must contain and control an unexpected problem with the objective to prevent a process upset. This can be logically carried out in the following steps:

1. Protect personnel. This need dominates all other actions.
2. Minimize damage to equipment.
3. Limit spills and releases to the environment.
4. Communicate effects of the problem immediately. Get help.
5. Notify all areas that may become involved. Use the assistance of the CCR operator and supervisor as communicators.

Once contained, control of the problem also can be accomplished in steps.

1. Work as a team to restore normal operating conditions.
2. Continuously update all involved areas on the status of the problem.
3. Bring in the technical specialists to start the problem-solving process as soon as personal risk is eliminated.

An activity that involves working with outside contractors or with central-plant support teams will experience fewer mistakes when planned and scheduled with input from the process operator of the area involved. The delivery of supplies, shipment of products, and schedules for extensive maintenance work are also activities that will benefit from the input of the process operator. Often a day process operator is assigned to coordinate complex activities with the shift teams. In any event, the process operator as troubleshooter keeps aware of future activity planned in his area and expects to be part of the planning effort, which is a sound strategy to prevent unexpected slip-ups.

The plant uses process operators to gather, record, and communicate timely and accurate production data that ranges from environmental reports to bills of lading. This work requires knowledge of the process to ensure accurate information, and networking with offsite contacts. These are examples where a day process operator is best positioned to do the job.

There is rarely a shift job that can be left unattended. The result is that a vacancy in a shift process-operator job is filled as soon as possible. To an experienced shift team, this is a climate for problem development, and a team effort is made to support the new member.

6.2 KNOW YOUR FELLOW TROUBLESHOOTERS

The process operator as troubleshooter must be familiar with the diversity of technical expertise that is included in the operations team. Process, technical, mechanical, electrical, instrument, environmental, and safety experts (among others) are there to assist in creating a safer, more productive operation. The process operator as troubleshooter takes the initiative to keep informed on the specific technical experts assigned to his area. This is his resource of those to call upon to help solve process problems. Keeping track of key technical personnel is not so easy since transfer of personnel is a regular occurrence in a plant organization.

The process operator as troubleshooter seeks the opportunity to meet newly assigned technical personnel to his area. He discusses with them a shared goal to improve operations by the prevention of the unexpected problem and technology available to better manage the intrinsic problems. Informal contacts offer an excellent opportunity to discuss possible process improvements and to explore working together to move them along. The core principle of a productive troubleshooting program is teamwork that fully exploits the technical expertise assigned to the operations team. The effective use of technical support is a learned skill that the competent process operator as troubleshooter has to master.

6.3 KNOW YOUR COMMUNICATIONS

The process operator transitions smoothly into a troubleshooter when he becomes the primary source of information on the effects of a process change that is accurately documented and communicated. The exclusive responsibility of the process operator as troubleshooter, after accomplishing containment and control of a process upset, is to supply essential technical information that the operation team needs to begin work on the analysis for cause. This point cannot be stressed too often. Without communication of timely and accurate information on the effects of a problem, effective troubleshooting is impossible. It is also a fact that more reliable information will come from the process operator, who knows his process technology. Only after root cause is deduced from factual information can the troubleshooting team, including the process operator, decide upon the action that will eliminate the cause and the problem.

The troubleshooting assistance on a process problem that the process operator provides comes in three forms:

1. An accurate description and documentation of the initial effects of a problem in correct, technical terms.
2. Continuously upgraded information on a problem as new information on effects surfaces.
3. Participation in and support of the team solution. Implementation of the action that will eliminate the cause of the problem.

Information provided on a process change is not edited, nor is any portion withheld by the process operator. There is often no apparent relation between an effect on a process change and who among the specialists has a need for that specific detail to solve the problem. The process operator as troubleshooter reports all the effects all the time with total objectivity. He fills this responsibility consistently by the documentation of precise, objective, and complete technical answers to the questions of where, when, what, and how on every process deviation from normal—including those resulting from human error.

The communications on a process problem must often involve several units. The process operator as troubleshooter needs input from other areas that have a process problem where the effects could have an impact on his area. This is especially true of an interruption in a process service, which must be communicated by the supplier to all dependant process areas immediately. This form of teamwork among shift teams is essential to minimize the number and severity of process upsets.

6.4 TEAMWORK RELIES UPON MUTUAL RESPECT

The capability to solve complex process problems is compromised unless the troubleshooting team respects the expertise of individual members. A troubleshooter that draws his conclusions without considering the merit of ideas introduced by others defeats the intent of the problem-solving process, which is to apply use of multiple, unbiased technical skills toward a consensus solution to a problem that is based on deductive reasoning.

The following suggestions will make working as a member of a troubleshooting team productive and certainly more enjoyable:

1. Arguments in support of a favorite cause are held in reserve until the validity of all the effects are confirmed.
2. Practice objectivity. Study the position presented by others that may conflict with your own. Seek common ground where technical agreement exists, and then deal with the contested differences.
3. Do not force the effects into a match with your favorite cause.
4. Look for weaknesses in the analysis of the effects and consider all probable causes carefully. The comparison of what is to what should be is an important exercise in the analysis of effects. The competent problem solver knows that unsupported causes will eventually fail the test of deductive reasoning.
5. Be open to arguments based on sound process technology even if the content is contrary to your expectations based on training and experience.

Every problem has a unique set of effects. The interpretation of these effects leads to a list of probable causes, which often requires input from several technical experts. The selection of the correct mix of technical expertise to successfully attack a problem is an important first step before any attempt is made to apply the problem-solving process. The objective of a troubleshooting team is to contain a mix of competency in operations, technology, and problem-solving ability that no individual working alone can be expected to muster.

6.5 KNOW HOW TO DETECT AND MANAGE PROCESS PROBLEMS

There is a continuous problem-solving effort directed at the detection and documentation of process information on intrinsic problems. The troubleshooter who uses a carefully planned plant patrol to monitor his unit for change will detect process upsets in the early stage of development. He can use plant patrol and the problem-solving process to accurately predict the timing on corrective action and better manage an intrinsic problem. New data on effects over a period of time produce progress toward finding root cause and eliminating the problem.

Inductive reasoning allows one to apply experience and training to solve a current problem when the solution fits a previous problem that occurred under similar circumstances with similar effects. Inductive reasoning is a valuable tool that the process operator uses daily to analyze a process change by comparison with a prior experience and in expert hands is a useful troubleshooting tool. Inductive reasoning is useful and most likely dominates with those who have no experience in team troubleshooting—that is what this course hopes to change over time. Traditional management of a process change by process operators relied on comparison to past incidents and this approach has worked for a long time. However, inductive reasoning may interfere with a search for root cause by reliance upon the similarities to a past problem, while distinctions are left unexplained. Not fully using one's experience and training is unreasonable, but knowledge of the process technology at work allows the process operator as troubleshooter to look beyond the obvious and see differences that may well produce a permanent solution.

The process operator plays a number of roles in managing a process problem. During the course of solving a problem, the process operator on the team may contribute as a production expert, be a technically astute eyewitness to an event, or have the responsibility to implement the process change needed to eliminate the cause of the problem. Each role supports the work of fellow troubleshooters who are not in the position or have the skills to do the job themselves. A good process operator follows two rules:

- Do not support any action based on guesswork or on unproven assumptions. Established procedures have worked up to this point and are far superior to a trial-and-error operation.
- Implement any solution that is supported by the operations team to the very best of your ability. Do this even if you have reservations about the merit of the solution. Be a team player. Hesitate only because of a serious process limitation or a safety concern.

6.6 KNOW THE PROBLEM-SOLVING PROCESS AND FAITHFULLY USE IT

Successful problem solving and troubleshooting requires a logical, unified approach among the problem solvers. The correct mix of technical expertise, experience, and skills focused on a well-defined problem is the combination needed to produce sound solutions. The problem-solving process methodically applies technical expertise to deduce what went wrong in the process technology using the effects as evidence. That is its strength. Teamwork is involved because of the number of technical fields involved. The process operator that has an education in process technology is invaluable as a problem solver in both supplying the effects and matching them with the problem.

A team leader is needed to keep the troubleshooting effort focused on a specific problem. If the team is permitted to continually expand the scope of the problem at hand, no resolution to a specific problem is likely. The troubleshooter on the team with the technical background to best explain the solution to the operations team is the logical leader. The process operator uses his training and on-the-job experience to support the guidance supplied by the team leader.

Successful problem solvers break complicated or convoluted process upsets down into smaller problems by working with each definable set of related effects. They assign a priority to each problem under consideration. The team leader can use this tactic effectively to concentrate effort on one specific problem at a time. The process operator helps the troubleshooting team to sort out effects based on his experience in matching effects to individual problems. Of course, this approach works best when the process operator in the field has knowledge of the process technology at work.

The problem-solving process requires patience and tolerance among the members of the problem-solving team so that deductive reasoning is allowed to lead to causes that best explain the effects. The viewpoint of each troubleshooter deserves full exposure during analysis and is rejected only for lack of technical support. There is no reason that inductive and deductive reasoning cannot lead to the same probable causes.

6.7 SUPPORT THE CONCLUSION AND ACTION TAKEN

To the process operator as troubleshooter, little seems to change. Proposed variations on old solutions to intrinsic problems are tried and tried again with little noticeable improvement in the performance of the unit. This situation does not discourage the dedicated troubleshooter.

He knows that there are many good reasons to implement change slowly in an established process that, for the most part, meets business objectives. No pragmatic operations team is eager to tinker with an acceptable unit performance no matter what the potential reward may be. That raises an important question: If the status quo is a profitable business strategy, why not stop all troubleshooting on intrinsic problems, accept the tolerable loss in utility, and make the best of the resources in hand? In other words, simply be content with plant performance as is and accept the performance in existing process technology?

Every business that accepts the status quo follows a faulted strategy that will eventually lead to loss of competitive edge and loss of profit. Shutdown of nonprofitable operations is inevitable. Even those small, incremental steps toward solutions to intrinsic problems improve plant performance over time. In fact, plant performance history will show that those small gains in handling intrinsic problems add up to significant increases in productivity over time.

Continuous improvement in productivity is a fundamental goal for the process operator as troubleshooter. An education in process technology is becoming a qualification for the job of troubleshooter. He understands that innovation must be given every chance to succeed.

One of the most productive means of support that the process operator as troubleshooter can provide is accurate and thorough feedback on the action taken to eliminate the cause of a problem. This is an essential communication that should follow both successful and unsuccessful implementation of solutions to process problems. Providing sound follow-up information to the troubleshooting team is indispensible to fully evaluate the results of the action taken and to plan any additional action. A well-written shift log, with all the details on a process change, often provides the information for planning the future troubleshooting effort.

Another positive way the process operator as troubleshooter can support a troubleshooting-team action is to contribute to changes in the safety and standard operating procedures that are needed to implement the change without risk. His skill and experience are invaluable assets in achieving a smooth transition into new or revised plant operations.

6.8 KEEP INFORMATION CURRENT

The need to gather information, analyze the effects, and deduce root cause as a workable problem-solving process has been stressed repeatedly as the basis for troubleshooting. The initial discovery and the effects reported are rarely a complete and accurate description of a problem. Even the initial action taken to contain a problem is wisely fine-tuned by troubleshooters to ensure results. The need to amend or correct information as it becomes necessary is an important contribution by the process operator as troubleshooter.

The serious troubleshooter continually reviews the available information on an intrinsic problem to keep his knowledge current. He communicates any additional information of his own immediately. This prevents false starts when analyzing the effects since the troubleshooting team will waste valuable time with incorrect or poorly defined effects. Prompt correction on information that keeps troubleshooters on track is a solid accomplishment of a competent troubleshooting organization.

6.9 WHO NEEDS TO KNOW?

The appearance of an unexpected problem is ever a possibility in a petrochemical plant. The motivated process operator as troubleshooter who detects an abnormal process change promptly reports what he has discovered. How does the process operator as troubleshooter decide on who needs to know about the first evidence of a process upset problem? He starts with the objective to bring the logical mix of troubleshooters to bear on the problem quickly. He does not raise unnecessary concerns among the general plant population who are in no way involved. The process operator as troubleshooter provides complete, unedited information to those assigned to the area with the mission to help troubleshoot problems. No data, including human error, are ever withheld or tempered, which could limit effectiveness of those with the direct responsibility for process performance. The process-control engineer is an example of the troubleshooter who must be fully informed on the unexpected problem immediately if he is to be effective. The process-control engineer uses his judgment to pass on information to other specialists on the technical team. He keeps the process operator informed on assistance he can expect.

As a rule, all other members on the operations team and plant staff should be informed of the unexpected problem through organizational lines, starting with the supervisor.

Information communicated beyond the needs of area troubleshooting teams on serious incidents is the responsibility of the shift supervisor. The process operator as troubleshooter fulfills his responsibility by fully providing information to those *who have a need to know*. Other inquiries are referred to the supervisor or his designate. This is one accepted method to control rumor about process-upset situations and incidents, yet get the information to those who have the responsibility to work on the problem. Rumors, as a rule, are an exaggeration of the facts, good and bad. Rumors will not help to solve the problem, and, if improperly spread on- and offsite, can needlessly worry plant neighbors and the general public.

6.10 SELLING THE SOLUTION

Even the well-thought-out solution and the correct, recommended action to eliminate the cause of a process problem have to meet safety and equipment-performance standards before any change in operations is considered. No change in standard operating conditions or procedures is likely to take place until production, mechanical, and technical teams are in agreement on the action to be taken and the expected results.

There are many good business reasons to gain broad approval before a change in operations in made. One convincing argument is the need for assurance that the best technical input has been used to rule out limitations in the equipment or a quirk in the process technology that could produce a hazardous condition. Even if a hazard possibility is remote, the change suggested has to be fully explored. A cross section of all troubleshooters should be involved in a significant process change for the following reasons:

1. Specialists are needed to expose and eliminate equipment limitations.
2. Standard operating procedures need to be revised to reflect the new or modified standard operating conditions.
3. Training of all shift teams may be needed.
4. Safety concerns and potential hazards are considered and eliminated. Any new or modified facilities are given a thorough inspection for safety and operability. A dummy run may be useful to check out equipment and procedures.
5. There may be environmental, quality, and other business concerns that have to be addressed.

The process operator as troubleshooter considers a change in operations from two significant viewpoints:

- Can the existing equipment perform safely and efficiently after the change is made?
- Does the existing instrumentation provide adequate information to follow and evaluate the effects of the change? The process operator as troubleshooter should not be involved in selling the solution to the problem unless he is fully convinced it can do not harm. His concurrence that the action is within his ability to implement is the reassurance needed by the troubleshooting team.

6.11 KEEP THE BOSS ON BOARD

Today's supervisor works with a number of shift teams who are assigned to operate several continuous processes that make up a plant process. Process operators may be asked to serve as members of troubleshooting teams from all disciplines in the operations team. This can create a workplace environment where the process-control engineer and other specialists are more involved in and better informed on specific process problems than the shift supervisor is.

However, the shift supervisor is involved in setting the priorities on the problems to be worked on and in the decision on who is capable to do the work. The supervisor has a need to be informed if there are changes in any process condition that could effect overall performance against production goals. The supervisor expects to be involved in any change in plans for plant activity, such as scheduled maintenance, and expects input from process operators on situations that involve personnel.

The process operator as troubleshooter is wise to keep his shift supervisor fully informed on any unusual activity in his area. The supervisor should be kept informed on the analysis and the action taken on unexpected problems. He has the right of first refusal on a proposed change in operations.

When an unexpected problem comes up, the process operator makes time to notifiy the shift supervisor at the earliest opportunity. Communicating to the supervisor through the CCR operator is common practice. The shift supervisor must be given the opportunity to bring in the technical assistance needed, and he expects to approve action being considered that is outside standard procedures prior to implementation.

It is good practice to fill in the shift supervisor on all information given to and obtained from fellow troubleshooters. This is another case where a pencil and notebook helps one recall important details. The shift supervisor must be in a position to communicate to plant staff about activity taking place within his area of responsibility. If a status report is requested to fill an offsite request for information, the shift supervisor is most likely the person onsite who fills the request.

The process operator sees the shift supervisor as an important participant and contributor to his troubleshooting effort. The shift supervisor must be in a position to pass on the activity of the shift team accurately and completely. His access to all plant resources in order to muster quick reaction to a problem is a valuable troubleshooting asset for the process operator.

6.12 WORK WITH PLANT STAFF

Members of the plant staff get out into the plant at unexpected times and frequencies. Usually, they are on a fact-finding mission to check out information they have been given. They may desire to get a better feel for operations in a specific area and to touch base with the personnel who work there. Often, they want to meet and chat with a new employee and see how well he is doing.

The process operator as troubleshooter should enjoy these contacts. He approaches and introduces himself to all visitors. Since he has a responsibility to keep the visitor out of harm's way, he points out areas that are off-limits and any potential hazards to avoid. He volunteers to escort the visitor (who must wear approved personal protective equipment), fill any specific need, or answer any questions.

In response to questions by visitors, the process operator as troubleshooter is frank and open, providing information that he is confident the visitor has a need to know. If unsure of his ground, he refers the questioner to his supervisor. It is also common courtesy to fill in the shift supervisor on the highlights of a visit by a member of the plant staff.

6.13 TRAINING

Knowledge of process technology and operational skills are basic tools the process operator as troubleshooter uses in his daily work. What is equally important is his ability to instruct and to train others. Assigned to use one's experience to train new personnel on the equipment and the procedures used to operate a facility is expected and should be considered a compliment. A less routine task, but one that also involves training, is working with contract personnel and mechanics who serve the unit. They most likely do not know the process. The process operator as troubleshooter has the responsibility to point out potential hazards and to review safety rules with all personnel before they begin work in his area. He must be satisfied that all work and the workers will proceed safely.

CLASS EXERCISE

EXERCISE 6.1 TROUBLESHOOTING IN THE TEAM ENVIRONMENT

The instructor will assign team members. He will fill the role of process control engineer.

Team Members	**Team Leader**
______________________	______________________
______________________	
______________________	
______________________	
______________________	

The operator that you relieved emphasized that he found a critical reactor feed pump with an overloaded motor. He put the spare into service and notified maintenance through the CCR operator.

After reading the logs and completing other necessary preparations for your shift, you start your first plant patrol. High on your list of priorities is the overheated motor problem. When you reach the location, the motor in question is gone. The spare pump seems to be running smoothly. However, two hours before the end of your shift, you carry out another patrol. The reactor feed pump motor is still missing, and the bearings on the spare pump are noisy and running much too hot. How long will they last?

The Facts

1. It takes one hour to shut down the reaction unit following normal procedures. The processes upstream and downstream also must shut down.
2. The processes upstream and downstream need two hours lead time to shut down following normal procedures (that is one-hour head start).

A crash shutdown on the reactor system takes three additional units down immediately. This would be the result of reactor feed-pump failure. A three-day cleanup operation and an additional day to start up would result.

Asides

1. The production supervisor calls three hours into the shift for a status report on the motor situation. He asserts that every pound of production is needed.
2. The process-control engineer expresses his concern about continuing to operate the spare. He reviews the consequences of pump failure. He suggests cutting feed rate until the motor is repaired. The operators upstream and downstream want to sweat it out.

Instructions

Each team member has 30 minutes to complete steps 1–3 below. The team has 20 minutes to complete steps 4 and 5.

1. Each class member as shift operator starts a timetable at the point he relieves the previous shift.
2. Following the teachings of this course, describe on the timetable in chronological order all the actions you would take to ensure good communications, teamwork, problem detection and management, and use of the problem-solving process.
3. Be sure that the actions you listed in step 2 handle this problem to your satisfaction.
4. Join your team as assigned by the instructor and prepare a master timetable with input from all team members. Is the team effort a better attack on the problem? Name the assistance the team used to carry out the actions taken. Did each team member have at least one good idea to contribute? Did the team come up with ideas that individuals did not come up with?
5. Would the troubleshooting efforts of a motivated process operator make a difference on the outcome of this situation? Would it help to have a technical team available to back you up?
6. Write up your timetable on steps 1–3. The team leader is to write up a summary table of the team effort.

CHAPTER

7

Prevention of Process Upsets

LEARNING OBJECTIVES

- See problem prevention as the ultimate accomplishment of good troubleshooting.
- Understand that problem prevention applies problem-solving techniques at a stage where a process upset can be prevented.
- Consider problem prevention the safer and orderly way to run a unit.
- See that problem prevention is the ultimate accomplishment of a process operator as troubleshooter.

KEY TERMS

Process upset

Procrastination

Process Upset
An operation unexpectedly carried outside standard operating conditions.

Process upsets are caused by faulty interaction among people, places, and things. Another truism is that preventing process upsets is far more productive than cleaning up after the fact. Problem prevention is the most difficult—yet the most rewarding—of the troubleshooter's accomplishments. The process operator is in the position to be a key participant in the effort to prevent process-related problems that lead to an upset. He puts his knowledge of process technology to use in the detection of an abnormal process condition in its early development and takes action to prevent development into a process upset.

Procrastination
Delay of a necessary activity until a future time.

Procrastination is a fatal fault in any problem-prevention effort. Effects on problems that the process operator as troubleshooter can prevent must be communicated immediately. A wait-and-see attitude consumes that small window of time available to analyze the effects and choose the best action, making problem prevention an option.

This is not to imply that any change in a process condition is a crisis or that taking action immediately will prevent a process upset. In fact, hasty action in itself may add to the problem. Often, hasty action based on incomplete information is more harmful than no action at all. An immediate communication of an abnormal process condition starts the process to prevent a problem through teamwork. Where corrective action is not obvious, the communication provides the appropriate technical experts with information that they can put to immediate use, while prevention of a process upset is still a viable possibility.

Production, mechanical, and technical troubleshooters work together on the operations team to prevent problems at all levels of responsibility. From his first day on the job, the process operator as a member of that team uses his knowledge of process technology in communicating accurate information on process conditions, which is put to use to prevent process problems.

The prevention of problems in new or modified facilities can take the form of posing and answering what if scenarios. Troubleshooters match potential problems with a variety of probable causes that can be eliminated before startup if recognized. The process operator as troubleshooter enjoys participation in discussions by teams assigned to propose and head off potential problems through the what if approach. Safety and emergency procedures, as well as process-control logic, are often explored and clarified through answers to simulated questions. For example, "What will happen if I do this?" is a fair question to ask and to answer before taking action to change a process condition.

Problem prevention requires a process operator as troubleshooter who accepts troubleshooting as a high priority in his job responsibilities. He is motivated toward early detection and immediate communication of any process abnormality. This requires a great deal of fortitude in petrochemcial plants that run well most of the time.

Effective problem prevention is centered around a process operator as troubleshooter who is constantly on the move to detect and report any questionable interaction among people, places, and things. This is optimum use of his troubleshooting skills and process knowledge.

7.1 CHANGE IS COMING TO PROBLEM PREVENTION

Staffing the operations team with competent technical personnel who have the responsibility to anticipate and prevent process problems has been the traditional approach to problem prevention. An education in process technology made problem prevention the domain of technical specialists.

This approach has two major drawbacks:

1. The technical specialists are on the plant site 25 percent of the available operating time.
2. The technical specialists are not positioned in the operating units in order to continuously detect problems and take corrective action to prevent a process upset.

The process operator is in the ideal position to troubleshoot process conditions throughout his shift. A process operator as troubleshooter with a working knowledge of process technology is ideally positioned to detect and report process abnormalities. A working knowledge of the process technology extends the skills of the process operator into analysis of effects through deductive reasoning. He can analyze effects of a problem in order to come up with a sound corrective action.

Full benefits of a troubleshooting program are advanced by a technically competent process operator who is in a position to anticipate and detect process problems, analyze effects, and take action that prevents process upsets. For complex situations, the process operator as troubleshooter is trained to solicit competent technical assistance to assist in analyzing the effects, deducing a cause, and deciding on the corrective action to eliminate the cause.

7.2 BASICS OF PROBLEM PREVENTION

Process systems that run continuously and are as complex as those in a petrochemical plant constantly undergo small variations in process conditions. The process operator as troubleshooter is trained to detect and pass judgment on a small change: Is it or is it not the first evidence of a process problem that could lead to a process upset? If the answer is "Yes," he knows there exists a root cause for the change, so he applies his technical knowledge and his experience to seek out and eliminate the cause. An action that eliminates the cause of a change early in the development of a problem is classic problem prevention and is expected from the shift team. However, there will be exceptions.

A small deviation from normal in a process condition can involve any combination of activities involving people, places, and things. Many preventable problems are initiated by human error or equipment malfunction, which could not have been anticipated. This is where the plant patrol is most productive as a routine search for signs of that unpredictable change.

The process operator as troubleshooter, with problem prevention in mind, understands there are climates for problem development (as shown in Chapter 4). When he expects problems under certain conditions, a major step toward discovery and prevention is in hand. He is not satisfied until any abnormal condition that he finds is corrected.

The problem-solving process is built upon complete and objective disclosure of effects for analysis through teamwork. The data provided by the process operator as troubleshooter in the shift log and through verbal exchange is an important source of information that shift and technical teams can analyze to anticipate and prevent problems.

A report on a change or effect that is important to one troubleshooter may not be significant to the problem-solving efforts of another. The process operator as troubleshooter recognizes the selective need for information by different technical specialists. He never censors the data that he reports, because he does not want to judge the value of any one effect to the individual members of the troubleshooting team.

The tools that the process operator needs for consistent detection and accurate communication of process changes are a working knowledge of the process technology, thorough training in process conditions, and a strong desire to troubleshoot. Knowledge and training will be more effective as the process operator as troubleshooter uses all the means of communication at his disposal to solicit assistance from the responsible technical, mechanical, and production troubleshooting teams.

The basis for problem prevention is the anticipation and search for small effects that indicate an unexpected problem may be developing, communicating those effects, and taking considered action. To fulfill that responsibility requires a great deal of perseverance as a means to find an abnormal condition early on in units that run well most of the time. Perseverance is a prevalent trait in the successful troubleshooter.

7.3 TRAIN FOR PROBLEM PREVENTION

Problem prevention works only through an operations team trained to discover abnormal conditions and to react quickly. It is well known to all troubleshooters that any delay in reacting to a process change is fatal to what may be an otherwise productive problem-prevention program. The process operator as troubleshooter firmly believes that every undersirable change has a cause that will worsen until defined and eliminated, which is one lesson that experience teaches. He also understands that competent technical assistance must be utilized.

The preparation for and completion of an effective plant patrol offered in Chapter 3, and the climates for problem development studied in Chapter 4, are at the forefront of problem-prevention training. Consistent use and refinement of the methods offered in those two chapters provide the process operator with the skills that are essential to a troubleshooter. Plant patrol is a proven system for discovery of an undersirable change in its early stages, and climates for problem development are common settings to anticipate and seek undesirable change. Doing both under the guidance of an experienced troubleshooter is a valuable part of an on-the-job training program.

7.4 SUMMARY OF THE PROBLEM-PREVENTION EFFORT

Effective problem prevention in a complex, technology-driven petrochemical plant is built around a highly organized, technical team effort. Troubleshooting teams dedicated to problem prevention in all plant activities do improve plant performance, which can be measured through advances in productivity. The troubleshooting effort can be divided into specific objectives.

Detection

1. The process operator is the key troubleshooter in the early detection of a process change, and plant patrol is his most valuable tool.
2. The process operator as troubleshooter networks continuously with technical and mechanical troubleshooters that share and compare information to solve process problems and prevent process upsets.
3. Early detection of problem effects prevents process upsets and is the reward of the troubleshooter who expects the unexpected.

Communications

Around-the-clock procedures set up for communication between shift and day personnel are well defined and should be consistently used. Most plants have continuous technical backup on call through the shift supervisor. Also, key technical personnel are made available during periods of high activity such as major shutdowns and startups, which are climates where problems are likely to occur. The shift supervisor can request the presence of technical specialists onsite at any time on a problem that requires their hands-on attention. Communications are centered around four points:

1. The process operator as troubleshooter is the key to the communication of accurate technical data, on any effect, to those who can assist in problem prevention. Skill and thorough training in communications is required.
2. The initial communication is with one's shift team, which includes the shift supervisor. The shift supervisor is in the best position to bring in outside assistance. The CCR operator can be of assistance by taking on the responsibility to alert related operations with a need to know.
3. The process-control engineer is a key technical assistant who must be kept informed. Any chemical or physical change in process conditions, change in process materials, quality problem, or question on equipment performance is within his area of responsibility. He can also assist in the communications with other technical specialists.
4. Field notes taken throughout the shift are invaluable memory enhancers that help keep technical information complete, organized, and accurate in a time frame.

Teamwork

The process operator as troubleshooter is, above all, part of a team in his effort to prevent problems. His working knowledge of process technology is applied to analyze a process change, which is followed by communication with the right troubleshooting assistance to choose the best corrective action. Going it alone in the early stages of problem development is not a strength but a serious flaw in the troubleshooting effort.

The process operator as troubleshooter treats every abnormal process condition as an early effect of a larger problem. There is no abnormality in a process condition insignificant enough to be ignored. He works with other troubleshooters until the condition is corrected or explained to his satisfaction. In fact, the process operator as troubleshooter makes a special effort to ensure that those involved in a troubleshooting effort are informed on every step through the final resolution on an abnormal process change. This builds credibility for future teamwork.

Who Needs to Know?

There is always some risk that communicating the effect of a potential problem in the early stages will turn out to be a false alarm. The objective is to communicate every potential problem to those with a need to know without unnecessarily alarming those who are not involved. Procrastination is far worse than being mistaken or overly zealous. A false alarm is of little importance compared to the reward in the prevention of a process upset or hazardous incident. One must be willing to be wrong, which will happen until one is fully trained and experienced on his job.

A decision by the process operator on those who need to know lies in the answer to two questions:

- Who can help define and confirm the existence of a potential problem?
- Who can contribute to analysis of the effects and expose probable cause?

Answers to the above questions place the information with the troubleshooters who can be of assistance and make a contribution to a sound solution. Unnecessarily alarming those who should be involved is a reasonable risk one takes to prevent a process upset.

Logs and Records

Day personnel depend upon shift logs and production records for information on shift activity twenty-four hours a day. The process operator is responsible for adequate detail and accuracy in all written communications to keep this important source of information available to needed assistance on materials, relief, and operations-team liaisons. A high level of effective communication is hard work and demands consistent effort by the process operator. Accurate and complete information from the field is an essential feature of a successful problem-prevention program that must include networking with the technical specialists.

It is difficult for the shift process operator to communicate consistently with day personnel. The chances are good that the troubleshooters who need to be involved in an unexpected problem will not be onsite when the early effects of a problem surface, analysis is made, and action is taken. Most troubleshooters, who have an interest in the problem, will follow up on information with the source identified in logs and records and seek more detail on effects from the source. Hindsight with the intent to prevent a repeat of the problem is good problem prevention.

The troubleshooters who use the data in logs and records are in a position to make a contribution to better team effort by follow–up and feedback to the source of information, who is likely the process operator. The process operator needs feedback on the content of the information he provides with suggestions on changes that improve communications. The process operator as troubleshooter may need to take the initiative and ask questions on how his information is put to use. The effort to become a better communicator is a career-long task.

Reports on team effort to solve a key problem is a necessary factor to keep troubleshooters focused. Reports are needed to confirm results and assure the information reaches the right technical specialist on the operations team. Specialists have an agenda of work, with set priorities they are expected to complete. Plant staff needs to support problem prevention as a key goal through assignment of technical resources. Reports justify this expense. Technical specialists are expected to respond to a request for assistance in reports.

Reports bring attention to a process problem and document the progress made toward a solution, and formally provide additional information as it becomes available. The process operator who asks questions on reports concerning problems in his area demonstrates another valuable trait of a troubleshooter.

Inaccurate, Ammended, and Additional Information

The need to report immediately the effects on a problem that is developing cannot be overemphasized, if prevention of process upsets is to be a meaningful goal of troubleshooting. There exists a high probability that the initial reports made under a time constraint will contain some degree of inaccurate and incomplete information. The innocent transfer of an occasional piece of inaccurate data should not deter the process operator from communicating quickly what he has in hand. Some confusion in the facts will not come as a surprise to any troubleshooting team who deals with the sudden appearance of unexpected problems. They understand that preventing process upsets demands quick action to explain a process change, and they will not fault a communication with the objective to get immediate assistance to prevent development into a more serious problem. Harm is done by a delay in bringing technical assistance into the problem-prevention effort because the process operator is reluctant to communicate less than complete information.

The solution is simple and direct: Communicate what is believed to be true and correct continuously, and add changes if necessary to all recipients of the original data.

Shift logs are a permanent record and sometimes become legal documents, yet a common occurrence is an entry that requires correction, amending, or clarification. The original record is corrected through a new entry in the current log with a reference to the previous data. *Never change the original entry.*

7.5 SUMMARY

Prevention of a process upset from an unexpected process change is the foremost goal of the operations team and, when successful, the foremost accomplishment in troubleshooting. The process operator with a working knowledge of process technology offers both anticipation and detection of an abnormal process condition in time to initiate a corrective response. On-the-job training provides the process operator with the necessary tools to use standard operating procedures to establish standard operating conditions. An education in process technology is the additional tool that the process operator needs to add detection and analysis of effects in order to correct a process change. The process operator as troubleshooter must continually expand and refine his knowledge in process technology; through observation and contact with technical experts, he expands his capability to detect, communicate, analyze, and participate on troubleshooting teams composed of all disciplines with the goal to prevent and solve process problems.

However well-trained and educated, the process operator as troubleshooter must apply himself to the consistent hard work needed to discover and communicate early effects of a process change. Problem prevention is simply not possible without early warning provided by a motivated process operator as troubleshooter. The prevention of a process upset, where equipment damage and hazards to personnel are often a possibility, is the ultimate reward of the motivated process operator as troubleshooter.

CHAPTER

8

Troubleshooting Exercises

Process troubleshooting is a skill that is broadened and fine-tuned with use and experience. The more numerous and varied the problems are that a troubleshooter works on, the better prepared he is to untangle the effects of the next one.

The exercises offered here place emphasis on troubleshooting as a problem-solving process that highlights the teamwork needed to discover, analyze, and draw conclusions from a given set of effects. They are broadly based and can be adapted to emphasize the points made in sections of the text.

The ability to communicate and document an undesirable change in a process condition and relate the change to the technology in use is a valuable combination of knowledge and skill, which can be practiced. The classroom exercises are based on three specific criteria:

1. Each exercise teaches the problem-solving process, with the "correct answers" to a process problem being of secondary importance.
2. Each exercise is designed with technical content that is within the comprehension of the novice process operator as troubleshooter who has completed the basic courses in equipment, instrumentation, and systems. If he is confused, he should refer to his previous studies as a refresher.
3. The figures in the exercises are not intended to be technical drawings of any type. Their purpose is to clue the student about more accurate and detailed drawings that may or may not be needed to complete the exercise. In plant life information and drawings of incidents on shift are rarely complete and lack detail that the operator must eventually supply.

The exercises are designed to use deductive reasoning to analyze effects, which is the foundation of the problem-solving process. The key is to organize the effects, from which probable cause can be deduced. Students can work as individuals, on teams, or both on any given exercise.

The successful student is expected to apply his studies in process technology, equipment, instrumentation, safety, and basic operations within the problem-solving process to guide his work on each exercise. The place to start is with a description of the problem derived from the effects communicated to the proper team of troubleshooters. This practice is routine troubleshooting. In a plant environment, providing information on effects would be the prime responsibility of the process operator as troubleshooter.

The exercises become progressively more complex as the novice troubleshooter becomes more adept in the use of the problem-solving process. Throughout the exercises, technical assistance is at hand from the instructor as process control engineer and should be sought out by the student. This imitates a typical, on-the-job team effort between the operator and the process-control engineer.

Teams of students are assigned to each exercise. Each team represents a typical mix of plant job assignments and responsibilities. A team composed of a student field operator, a student day operator, and a student CCR operator with a supervisor for backup would be one possible team makeup. One team member is selected as leader and moderator. His reponsibility is to keep the team focused on the exercise instruction. He is expected to lead the class discussion on the efforts of his team in a class presentation.

Some exercises are presented with three different sets of effects on one process system, which creates three different problems. Each team can work on all three problems, or three different teams can be assigned to work on a separate problem. The instructor may place time restraints on team activity to create a sense of urgency.

Each team member prepares a summary report on the exercise using the worksheet format (as shown in Figure 8–1). He is encouraged to spell out his contribution to the deductive-reasoning process that provided the team solution.

The instructor appoints himself process control engineer and serves as the technical advisor to the student troubleshooting teams. This is a source of technical assistance that is readily available among technical experts in the workplace, and its use is essential to good performance by the process operator as troubleshooter.

As requested by the instructor, each team will present an oral summary of the assigned exercise to the class. Teams can expect to be questioned and

NAME: ______________________________ DATE: ______________

DISCOVERY

PROBLEM DESCRIPTION

ANALYSIS

CONCLUSION

ACTION

Figure 8–1 Troubleshooting Exercise Worksheet

are expected to provide sound technical answers to support their arguments. The oral presentation describes the problem-solving process used by each team to analyze the effects and support a list of probable causes. This is a common plant setting where troubleshooters get together and review progress as they seek a solution to a problem.

8.1 DISCOVERY

Every exercise starts by each student troubleshooter describing the action to take that assures his safety and the safety of others. The team approves the containment and control actions taken immediately upon discovery of the problem. The stage is set for the troubleshooting activity through teamwork that follows.

The student troubleshooting team thoroughly takes notes on the effects needed to reconstruct in time the sequence of events prior to the action taken after initial discovery in the field. The student process operators should communicate the effects immediately on a problem to fellow troubleshooters who need to know. This is normally carried out through the CCR operator.

8.2 DOCUMENTATION

After the process is under control, it is time to document the effects. No detail can be considered unimportant at this point. The student troubleshooter must answer the following questions:

Where? When? What? How?

These steps will help:

1. Document all the effects of the problem. A sketch may help.
2. Copy those who need to know.
3. Describe the immediate action that was taken, and who was involved.
4. Note any background information that may contribute to a better understanding of the effects.

8.3 ANALYSIS

The effects are studied with the intent to develop a concise description of the problem. Distinctions are especially important points. What it is? What it is not? Also, what information, if any, is missing?

At this stage, the need for additional information may surface and be noted. The process-control engineer (instructor) may well be asked to clarify a technical point or to provide additional information. The tendency to jump to a conclusion is resisted. The team leader and instructor should insist that the team study the effects (or ask for more) until the basic questions are answered and the problem can be concisely described in writing. What it is? What it is not? What should be?

It helps if one:

1. Accurately summarizes the effects in a timed sequence from start to finish.
2. Notes additional information needed and suggests how it can be obtained.

8.4 LIST PROBABLE CAUSES AND DRAW CONCLUSIONS

Discuss and make a list of all possible causes that explain one or more effects. Remember that causes and effects are often interchangeable in a timed sequence. When the list of causes is complete, start on conclusions. The team can conclude at this point that the problem is unique and not likely to repeat. The action taken upon discovery is adequate to prevent reoccurrence, no additional tests and/or information is needed, and no additional action is needed to eliminate the problem. Possibilities to reduce the chance of reoccurrence include additional training, process modification, or revised procedures. Follow these steps:

1. Apply deductive reasoning to develop probable causes from the effects.
2. Apply deductive reasoning to reduce the list of causes to the most probable. This is a test of process knowledge and judgment.
3. Choose the probable cause that explains the most effects.
4. Summarize the problem-solving process using the form provided in Figure 8–1.

8.5 ACTION

Action is the change made to eliminate a cause. In an industrial setting process change requires consensus and follows an organized plan for approval. It is important that the process operator as troubleshooter be a full participant in the decision on what action to take that will eliminate the cause. It is also important that all others with operational responsibility be informed of any action and how it may affect them. Two steps lead into approval of a process change:

1. Recommend action that eliminates the cause. Write it up in a form that will educate and sell the solution to the operations team.
2. Summarize all action that was taken upon discovery of the problem and the results obtained. This is strong support for a permanent change.
3. Communicate any proposed action to all the plant areas that may be impacted. This prevents problems before initial action is taken.

EXERCISE 8.1 LEAK ON A FLANGED CONNECTION

Team Members	**Team Leader**
____________________	____________________
____________________	
____________________	
____________________	

A leak on a flanged connection is one of the most common equipment problems in a petrochemical plant. It can happen anywhere, at any time, on any process material, and under any conditions.

The cause(s) can be obvious or well disguised. That makes it an ideal problem to practice troubleshooting using the problem-solving process.

You will work in teams to obtain experience in teamwork and communications.

The objective of the exercise is to correctly apply the troubleshooting principles and process technology through use of the problem-solving process.

Each team will be graded on its approach to the problem and effective teamwork. Finding the correct cause is important but the use of good troubleshooting technique is more so.

Each member of each team is to submit a written summary on the exercise using the worksheet format in Figure 8–1–1. You may use the back or additional sheets if necessary. Write the summary as you would a shift log.

Use your imagination and be creative in your approach to this basic process problem.

Table 8–1–1
Leak Effects

Problem	**I**	**II**	**III**
Location of connection	Pipe to pipe	Pipe to vessel	Pipe to tank
Line size	1″	4″	10″
Material	Steel	Stainless steel	Steel
Operating Temp. (°C)	25	150	50
Operating pressure (psig)	10	1,000	1 to 3
Line insulated	No	Yes	Yes
Line traced	Yes	No	No
Elevation	Below head level	8′ above floor level	Ground level in a tank farm

Table 8–1–2
Process Material

Problem	**I**	**II**	**III**
Estimated leakage (PPH)	10	5	20
Toxicity	High	Low	Moderate if ingested
Flammable	Yes	No	Yes
Boiling point (°C)	70	150	200
Freezing point (°C)	10	50	0
Phase	Liquid	Mixture (gas/liquid)	Liquid
Leak description	Spray	Hot drip	Drip
Corrosive			
Steel	Yes	No	Slightly
Stainless	Yes	No	Slightly

Table 8–1–3
Supporting Information

	I	II	III
Time: Wed.	12:00 noon	12:00 midnight	Not known
Atmos. temp. (°C)	5	25	20
Plant status	Shutdown in progress	Normal operation	Startup (tank is full at 80% level)
Location	Process building	Process building	Tank farm
History of leaks	Yes	No	No
Recent maintenance	Not known	Not known	Not known

Table 8–1–4
Possible Causes of the Leak

Gasket failure	Cocked flange
Incorrect gasket	Liquid expansion
Improper bolt tension	Thermal cycles
Pressure surge	Temperature spike

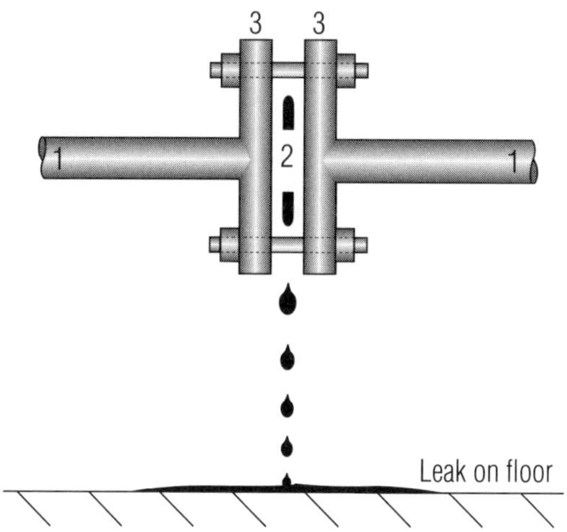

Key:
1. Pipe
2. Gasket
3. Flanges

Figure 8–1–1
The Pipe Connection

Figure 8–1–2
Flanged Connections

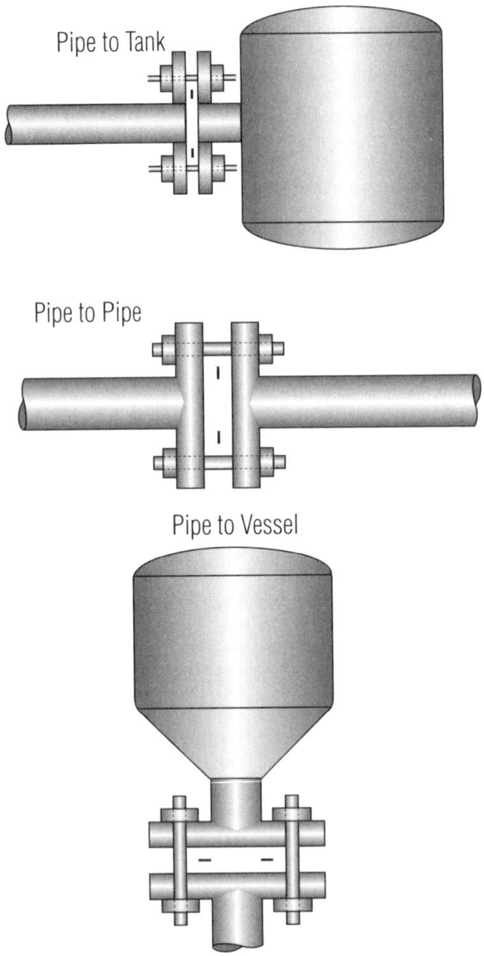

Instructions

1. The effects you will work with are shown in Table 8–1–1.
2. A specific problem is assigned to each team. See Tables 8–1–1 and 8–1–2; Problems I, II, and III; and Figures 8–1–1 and 8–1–2.
3. What immediate action would you take upon discovery? Enter your thoughts under Discovery on the worksheet.
4. Study the supporting information.
5. Name the individual troubleshooters from the operations team who would participate in a plant setting.
6. List the probable causes that relate to the effects.

7. Do you need more information? If so, request it from the process-control engineer (instructor).
8. Decide on the most likely cause. Explain your choice.
9. What action do you recommend?
10. How would you test the action?
11. Have you used all the operations team resources available to you in the solution to this problem?

EXERCISE 8.2 THE CENTRIFUGAL PUMP

Team Members	**Team Leader**
______________________	______________________
______________________	
______________________	
______________________	

The centrifugal pump is one of the most dependable pieces of plant equipment used to transfer liquids. Performance of pumps has a significant effect on plant utility. The student process operator as troubleshooter makes sure that his or her pumps are performing properly at all times through routine patrol with the intent to uncover any sign of malfunction.

Comparison of the performance of a centrifugal pump against specifications is easily made by use of performance data chart (Table 8B.6). Every pump has a specific set of performance parameters that includes net positive suction head (NPSH), discharge head, flow rate, required horsepower, impeller size, and rotational speed. The process operator as troubleshooter checks *any* change made to a pump against the expected performance curves to ensure that process demands are met.

The most common process problems with a centrifugal pump, other than mechanical failure, usually involve an inadequate NPSH, a change in the material being pumped, a change in process temperature, modified piping, or a change in process flow rate. All can result in overload, cavitation, or vibration, which are symptoms of a pump in trouble.

A common mechanical problem with the centrifugal pump often involves the shaft seal to the pump casing. A variety of seal designs are used to prevent leakage of process materials to the atmosphere. Early detection of the signs of seal failure, which usually starts with a very small leak, is an important troubleshooting discovery.

Table 8–2–1
Effects

	A	B	C
Discharge pressure (psig)	13	Fluctuating 20 ± 5	25
Suction pressure (see Figure 8–2–1)			
Y in feet	12	2	10
X in feet	3	3	3
Flow (mpph)	200	150	100
Rattling noise	No	Intermittent	Yes
Vibration	Yes	No	No
Motor shutdown	Yes	No	No
Motor overheating	Yes	No	No
Specific gravity	1.10	0.90	0.80

Table 8–2–2
Supporting Information

1. Piping arrangement — Figure 8–2–1
2. Pump schematic — Figure 8–2–2
3. *Pump size 4 × 3
   Pump speed (rpm) — 1,750
4. $P_T = 20$ in. of H_2O — Ts = 25°C (from Figure 8–2–1)
5. $\text{Flow, Mlb/hr} = \dfrac{(8.33)(60)(\text{gpm})(\text{spec. grav.})}{1000}$
6. Motor horsepower is proportional to impeller diameter to the third power.
7. $\text{Total head} = \dfrac{2.31}{\text{Spec. grav.}} \times \text{discharge pressure, psig}$

Type seal	Soft packing	Mechanical	Mechanical
8. Recent maintenance	Yes	No	No
9. Impeller size	9″	8″	7″
10. Motor size (hp)	5	7.5	7.5

*Pump size is designated by suction on discharge connections. 4 × 3 would be a 4-inch suction and a 3-inch discharge.

Table 8–2–3
Useful Equations

1. Net Positive Suction Head (NPSH)

$NPSH = Ps + Hv - Vp$
Ps = suction gage pressure in feet of head
Hv = velocity head in feet of head
Vp = vapor pressure in feet of head
Head = feet of fluid pumped

2. $$\textbf{Feet of Fluid} = (\text{psig})\frac{(33.4)}{14.7}(\text{sp. grav.})$$

33.4 = feet of water in one atmosphere of pressure
14.7 = psi in one atmosphere of pressure
sp. grav. = specific gravity of fluid being pumped

3. $\mathbf{Hv = 0.0155\ V^2}$

0.0155 = gravity constant $\frac{1}{2}g$
V = velocity of fluid flow in ft/second

$$V = \frac{\text{flow of fluid in ft}^3\text{/second}}{\text{cross-sectional area in ft}^2\ (Cs)}$$

$$Cs = \Pi r^2 \text{ r in inches}$$
$$= (3.14)\left(\frac{\text{r in inches}}{12}\right)^2$$

Cs = cross-sectional area of pipe (ft^2)
r = pipe radius in inches
π = Pi = 3.14

4. $$\textbf{Flow in ft}^3\textbf{/second} = \frac{\text{gpm, gal./min.}}{(60\text{ sec/min})(7.48\text{ gal./ft}^3)}$$
$$= \frac{\text{pph, lb/hr}}{(62.4\text{ lb/ft}^3)(3600\text{ sec/hr})(\text{sp. grav.})}$$

5. Total Discharge Head

A. Total discharge head $\mathbf{H_D = (P_D) + (H_v)}$

P_D = discharge press (ft)
H_v = velocity head (ft)
Calculate P_D (see ft. of fluid)
Calculate H_v (See NPSH calculation)

B. Total head developed by pump

$H_T = H_D - NPSH$

Table 8–2–4
Centrifugal Pump Possible Problems, Causes, and Solutions

POSSIBLE PROBLEMS		
Seal leak	Excessive seal wear	Motor vibration
Low discharge pressure	Motor shutdown	Pump surge
Pump vibration	Low flow	Vapor lock
Cavitation		
POSSIBLE CAUSES		
Worn wear ring	Pump vibration	Poor insulation
Low suction head	Impeller size	Oversized pump
Cracked seal	Motor overload	High suction temperature
Coupling failure	Worn seal	Viscosity
Low NPSH	Suction-line restriction	Pump not vented on startup
Internal circulation	Worn bearings	
POSSIBLE SOLUTIONS		
Lower temperature	Shut down and clear	Reduce NPSH requirement
Agitate tank (feed)	Reduce horsepower	Shut down and vent
Increase tank level	Reduce speed	

Table 8–2–5
Centrifugal Pump Troubleshooting Aids

Problem	Possible Cause	Solution
Specific gravity of product is about to change	• Different product composition • Starved suction	• Leave alone; will not affect pump capacity • Pinch down discharge value on pump
Excessive vibration	• Formation of vapor pockets • Pump misaligned	• Vent pump to re-establish full capacity • Shutdown and have realigned
	• Rotor out of balance; usually intermittent	• Remove rotating element, check impeller; if passages clogged, remove foreign material; if impeller is damaged, a new one will need to be installed
	• Shaft bent • Foundation and pump not connected • Driver vibrating	• Remove shaft • Secure pump to foundation • Disconnect coupling and check driver
	• Control valve installed improperly	• Relocate valve
Pump is cavitating	• High suction temperature • Viscosity • Low suction pressure • Restriction in line • Too much horsepower • Pump speed too high • Too-high NPSH standard • Small suction	• Lower temperature • Agitate suction tank • Increase level or pressure • Shut down and clear • Smaller horsepower • Lower rpms • Lower NPSH requirements • Larger diameter suction line
Pump vapor is locked	• Pump not vented prior to startup	• Shut down and bleed off

Table 8–2–6 Pump Performance Data

Impellar Size	Horsepower	Eff; %	NPSH, ft	Total Head, ft	Flow, gpm
7″	<5	<50	3	47	100
	<5	65	5	42	200
	<5	72	6	35	300
	<5	55	8	23	400
8″	<5	<50	3	62	100
	<5	68	5	57	200
	5	78	6	50	300
	$5\frac{1}{2}$	73	8	39	400
9″	5	<50	3	78	100
	6	60	5	76	200
	7	78	6	70	300
	$7\frac{1}{2}$	80	8	62	400
	$7\frac{1}{2}$	68	12	47	500

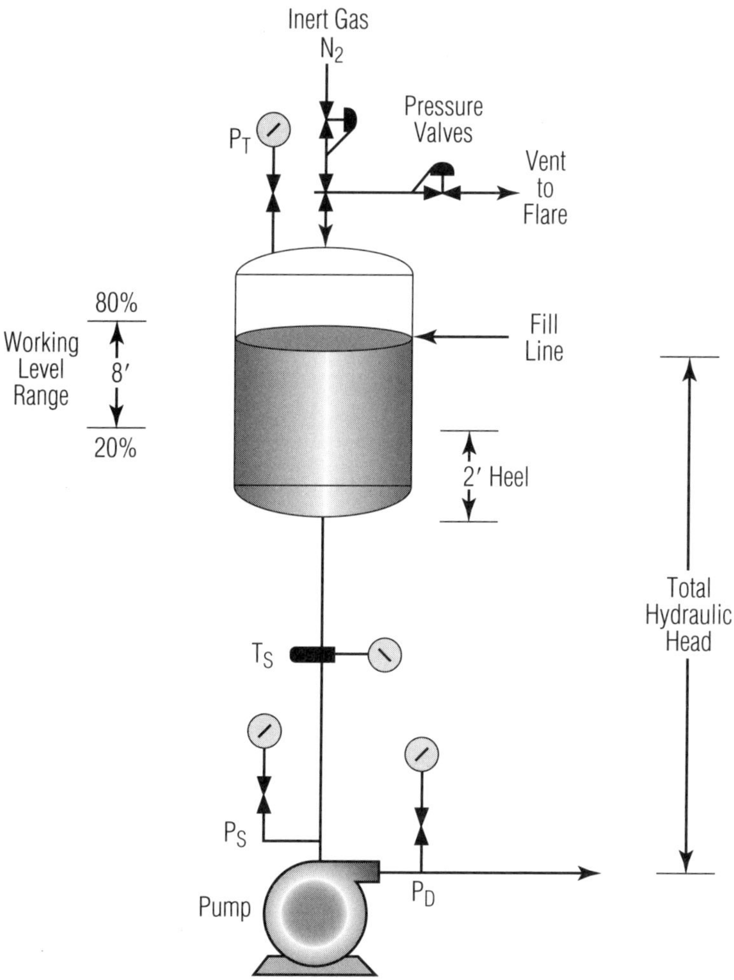

Figure 8–2–1 Piping Arrangement of a Typical Pump to Show Importance of Net Positive Suction Head

Figure 8–2–2
Typical Pump Schematic to Show Major Components

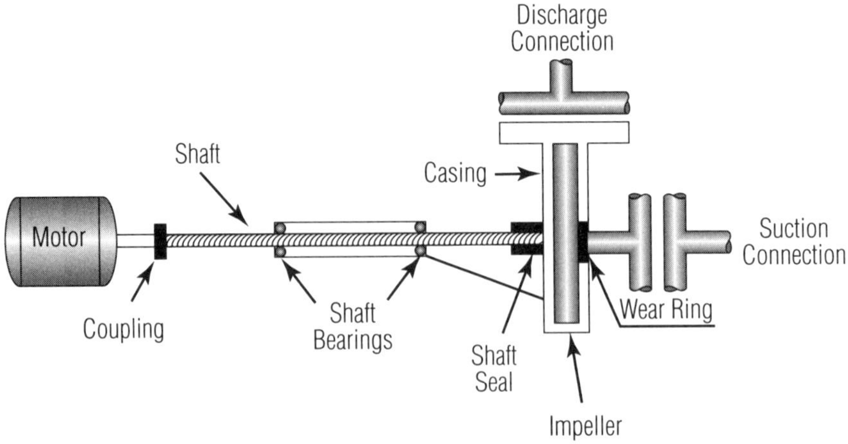

Instructions

1. The effects on the problem you will work with are listed in Table 8–2–1.
2. Process supporting information is listed in Table 8–2–2.
3. Use Figures 8–2–1 and 8–2–2 and Table 8–2–6 to define and make the necessary calculations.
4. Apply the techniques in the Discovery section to write a statement defining the problem.
5. Do you need operations team assistance? Name the operations team members you want on the troubleshooting team.
6. From an analysis of the effects, identify and list probable causes that explain the effects.
7. Do you need more information? If so, describe your needs to the process-control engineer (instructor).
8. Decide on the most likely cause. Explain your choice.
9. What action do you recommend?
10. How would you test the recommended action?
11. What operations team resources did you use?
12. Complete the worksheet.

Basic Calculations Procedure

Analysis of all data starts with consistent use of terms, symbols, and units of measurement. No troubleshooting team can function until all the members are working from a consistent set of information.

1. In Table 8–2–1, the flows are given in mpph. The performance data (Table 8–2–6) are in gpm. Convert mpph to gpm (Table 8–2–2, equation 6).
2. Calculating the net positive suction head (NPSH) is one key to analyzing pump performance. In Table 8–2–6, NPSH is in feet of fluid, and so is the total head (discharge pressure). Most plant pressure gauges will measure pressure in psig. Using equation 8 in Table 8–2–2 and the equations in Table 8–2–3, calculate NPSH and total head in feet. Remember, specific gravity is an important factor.

■ EXERCISE 8.3 THE REFRIGERATION UNIT

Team Members	**Team Leader**
____________________	____________________
____________________	
____________________	
____________________	

Refrigeration is a common industrial process found in petrochemical plants. Low-molecular-weight hydrocarbons are processed at temperatures and pressures below the range of natural coolants like surface or cooling-tower water.

A single refrigeration unit is often used to supply coolant to several process users because of the economy in size. Brine (salt solution) is a common fluid used as the circulating coolant to remove heat from the process. Brine is inexpensive and has a high heat capacity. Corrosion is the most common problem.

It is not uncommon for several areas and process operators to be involved if there is a problem with a refrigeration system. Good teamwork and troubleshooting skills are needed to limit the effects of a refrigeration problem by quickly finding a root cause that can be eliminated before a process upset occurs. In this exercise, a student CCR operator, a student area operator, and the student operating the refrigeration unit are involved with different effects from the same refrigeration unit.

Your assignment is to work together as a team to find the most probable cause and recommend the action needed to correct it.

The Situation

A. The CCR operator has informed the area process operator of a low compressor suction pressure alarm on his refrigeration unit. (Figure 8–3–1, 8–3–2)

B. The process operator in another process building, which uses refrigerated brine from the refrigeration unit above, reports that the temperature of the brine supply is up to 54°F. (Figure 8–3–1, 8–3–2)

C. The area process operator discovers that the cooling water flow on the condenser in the refrigeration unit in question is approaching maximum in order to maintain the desired refrigerant-condensate temperature. (Figure 8–3–1, 8–3–2)

Table 8–3–1 Effects Identified on Figure 8–3–1

	Standard Operating Conditions	Actual Conditions
T1	85°F	85°F
P2	95 psig	95 psig
L2	50% ± 10	70
TV2	50%	100%
L1	20%–80%	0
T2	50% ± 5	53°F
T4	60°F ± 5°	61°F
T3	35°F ± 5°	42°F
P1	32 psig ±	*20 psig
TV1	50%	90%
FW	100–200 gpm	150 gpm
T5	90°F	90°F
T6	70°F	70°F
S1	800–1200 rpm	1200 rpm

Field Readings

L1	Evaporator refrigerant level (%)	
P1	Evaporator refrigerant press. (psig)	
T3	Evaporator refrigerant temp. (°F)	
T4	Brine inlet temp. (°F)	
T2	Brine exit temp. (°F)	*49(**50 ± 5)
S1	Compressor speed (rpm)	*1000(**800–1200)
	Compressor discharge pres. (psig)	
	Compressor discharge temp. (°F)	
T1	Condenser exit temp. (°F)	*85(**80 ± 5)
P2	Condenser cooling water flow (gpm)	*94(**100–200)
L2	Reservoir level (%)	*70(**50 ± 10)

*Last logged readings by field operator.
**Acceptable range.

Instructions

1. The actual conditions that you will work with are listed in Table 8–3–1.
2. Standard operating conditions are also listed in Table 8–3–1.
3. Supporting information is listed in Tables 8–3–2, 3, and 4.
4. Write a statement defining the problem.
5. Do you need assistance? Choose the troubleshooting team.
6. From an analysis of the actual conditions (effects), identify the possible causes. Can you explain the effects?
7. Do you need more information?
8. Choose the most likely cause. Explain your choice.
9. What action do you recommend?
10. How would you test the recommended action?
11. Turn in your write-up to the instructor.

Table 8–3–2
Supporting Information

TIMETABLE

	Situation A	Situation B	Situation C
Time	6:00 A.M.	12:00 P.M.	12:00 A.M.
Amb. temp. (°F)	70	90	40
Plant operations are normal			

PHYSICAL PROPERTIES

Component: Brine	Property Value
Wt. % $CaCl_2$	9.2
Freezing point, °F	22.5
Specific gravity (60°F)	1.1
Enthalpy, Btu/lb	
40°F	214.4
50°F	223.1
60°F	231.8
70°F	240.6

Freon 12 Refrigerant

Vapor pressure	Table 8–3–6
Heat of vaporization	75 Btu/lb
Heat capacity (vapor)	0.2 Btu/lb

Table 8–3–3
Physical Constants

1 standard ton	12,000 Btu/hr
1 standard ton	4.7 horsepower
1 standard ton	3.51 kilowatts
1 horsepower	746 watts

Table 8–3–4
Safety Features

Safety Device	Failure Setting
Evaporator relief valve setting (psi)	70
Reservoir relief valve setting (psi)	150
Compressor overspeed trip (rpm)	1,250
Compressor low suction pressure alarm (psig)	20

Table 8–3–5
Problems

Compressor shutdown
Reservoir relief valve failure
High condenser exit temperature
Evaporator relief valve failure
Low brine exit temperature
High condenser cooling water flow

Table 8–3–6
Freon 12 Data

Temperature	Vapor Pressure (psig)
0	9
50	43
100	100
150	220

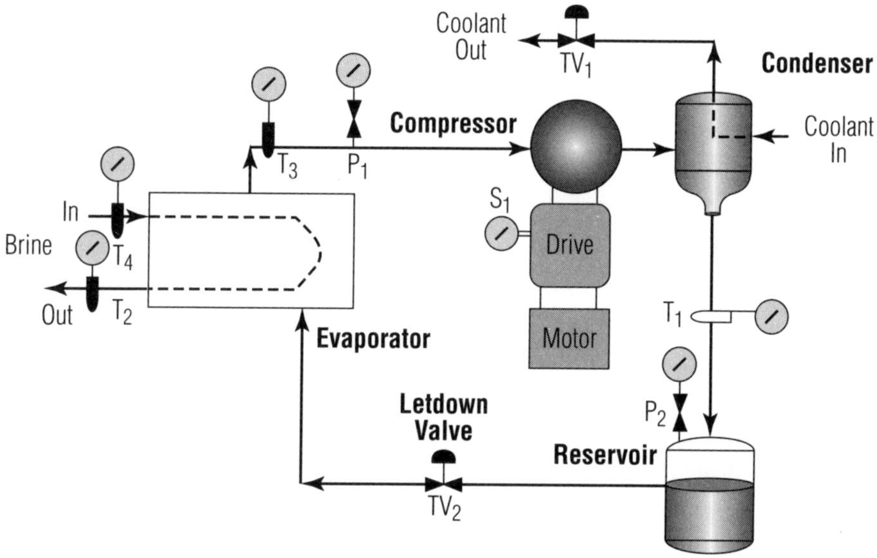

Figure 8–3–1
Refrigeration Vapor Compression

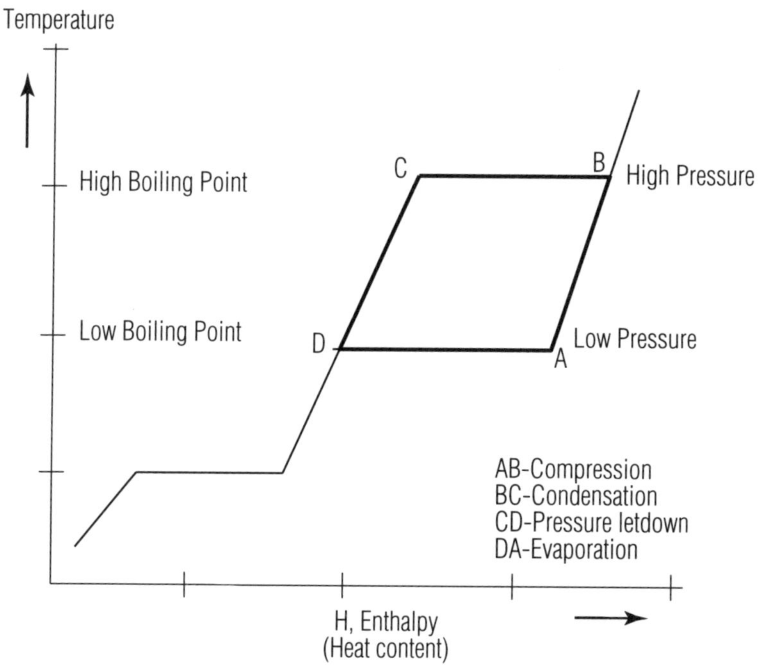

Figure 8–3–2
Typical Refrigeration Cycle Vapor Compression Cycle

■ EXERCISE 8.4 ION EXCHANGE

Ion exchange is a common process used to remove a low concentration of impurities from liquids. Ion exchange is less expensive than distillation in the demineralization of water for process use.

Ion exchange is a replacement of one chemical ion for another.

Most demineralization units are fully automated, which provides push-button operation to initiate the loading and regeneration cycles. However, this does not reduce operator responsibility to monitor and troubleshoot the process

Instructions

1. An ion-exchange unit is being used to demineralize soft water. Sulfuric acid (H_2SO_4) is used to regenerate the cation beds, and sodium hydroxide (NaOH) is used to regenerate the anion beds.
2. The process has five basic steps, as shown in Figure 8–4–1. Typical piping is shown in Figure 8–4–2. Figures 8–4–3 and 8–4–4 are provided for detail on the process.
3. One cation and one anion bed are in service at all times. The offline beds are either being regenerated or put in a standby mode.
4. This is a package unit and is serviced by a contractor. The resin is replaced annually. It is purchased by the contractor and shipped directly to the plant site. The contractor provides the resources to replace a bed, and the work is completed under the supervision of the contractor.
5. Using the assigned effects, evaluate the status of beds.
6. Write a description of any potential or actual problems.
7. Make a list of possible causes.

PROCEDURES

Draw a theoretical loading curve for the cation and anion beds.
Draw a theoretical regeneration curve for the cation and anion beds.
Draw an actual loading curve for the cation and anion beds.
Draw an actual regeneration curve for the cation and anion beds.
Match the effects with the loading curves.

Table 8–4–1
Standard Operating Conditions as Shown on Figure 8–4–1

Process Condition	Setting
Process Flow (F_1)	84 gpm
Regeneration flow	
F_2	20 gpm
F_3	10 gpm
P_2	100 psig
P_3	30 psig
P_4	80 psig
P_5	25 psig

Table 8–4–2
Basic Data

Anion bed loading (theoretical)	
OH^- as moles per ton of resin	1.18
$SO_4^{=}$ as moles per ton of resin	0.59
Cation bed loading (theoretical)	
H^+ as moles per ton of resin	0.20
Na^+ as moles per ton of resin	0.20
Regeneration solutions	
H_2SO_4	2.0 wt. %
NaOH	2.0 wt. %
Anion bed loading (expected)	
OH^- as moles per ton of resin	1.00
$SO_4^{=}$ as moles per ton of resin (new bed)	0.50
$SO_4^{=}$ as moles per ton of resin (spent bed)	0.40
Cation bed loading (expected)	
H^+ as moles per ton of resin	0.18
Na^+ as moles per ton of resin (new bed)	0.18
Na^+ as moles per ton of resin (spent bed)	0.15
Feed water impurity concentration	
Na_2SO_4	200 ppm
Demineralized water impurity (product)	
Na_2SO_4	10 ppm
Bed weight and life	
Cation exchange bed weight (tons dry)	12.5
Anion exchange bed weight (tons dry)	12.5
Expected life of bed (mm lb.)	100.0

Table 8–4–3
Standard Operating Conditions: Process

Feed rate (gpm)	10
Feed concentration (ppm)	
Na^+	200
SO_4^+	200
H^+	0
DH	0
Make concentration (ppm)	
Na^+	10
$SO_4^=$	10
H^+	0
DH^-	0
Inlet pressure (psig)	50
Exit pressure (psig)	30
Load cycle (hr)	168

Table 8–4–4
Standard Operating Conditions: Regeneration

Cation Bed	
H_2SO_4 solution flow (gpm)	*
Regeneration cycle (hr)	12
Anion Bed	
NaDH solution flow (gpm)	*
Regeneration cycle (hr)	

*By student troubleshooters.

Table 8–4–5
Actual Conditions: Process

SITUATION	**A**	**B**	**C**
Inlet flow (gpm)	10	7	15
Inlet conc. (ppm) Na_2SO_4	200	200	200
Exit conc. (ppm) Na^+, ppm	5	10	40
$SO_4^{=}$, ppm	5	10	40
Inlet press. (psig)			
Exit press. (psig)			
Total throughout (Mm lb)			
Regeneration			
*Inlet flow (gpm) NaOH			
H_2SO_4			
**Inlet press. (psig)			
Exit press. (psig)			

* Provided by student troubleshooter.
**Estimated by student troubleshooter.

Figure 8–4–1
Five-Step Regeneration

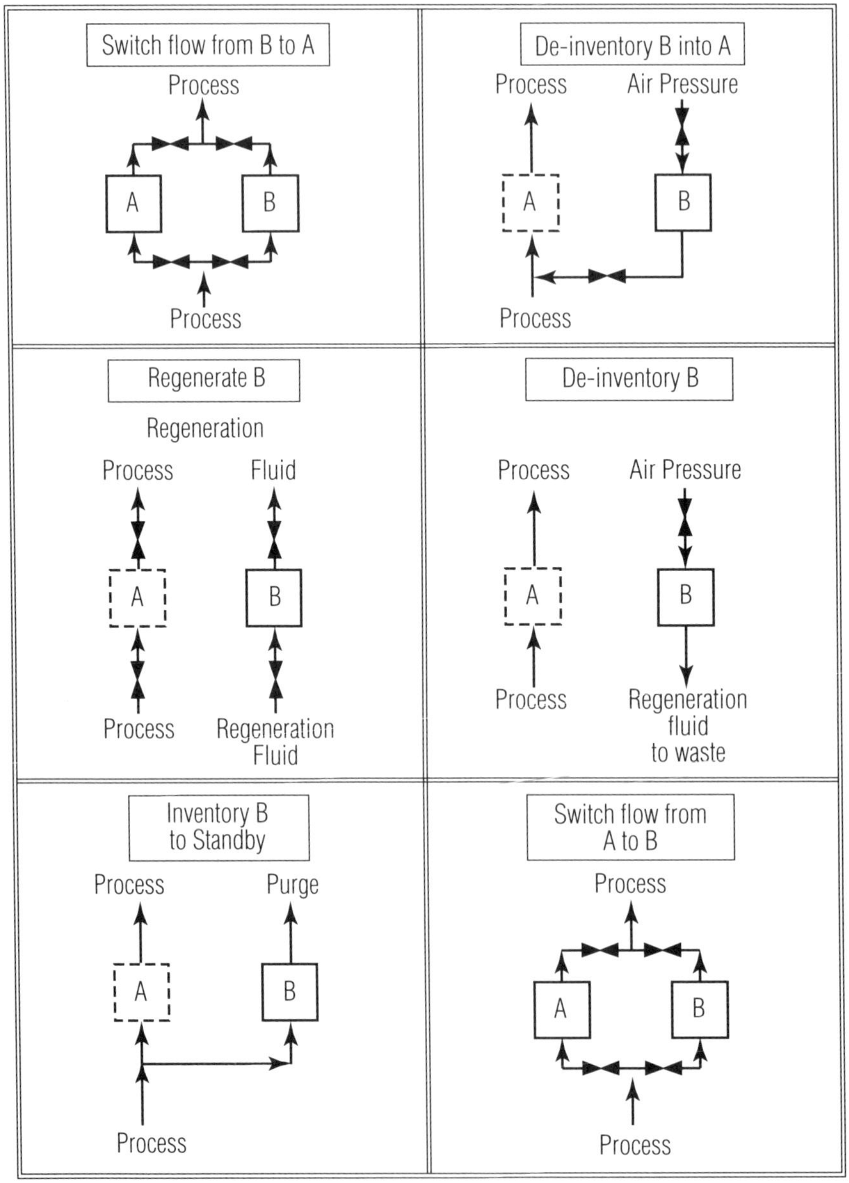

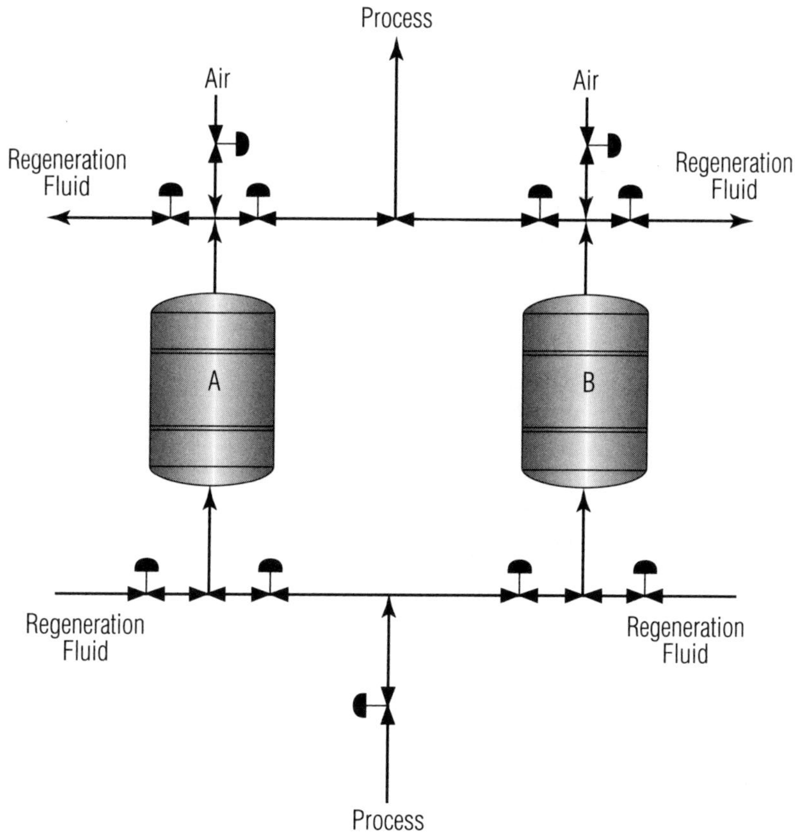

Figure 8–4–2
Ion Exchange Valve Arrangement

Figure 8–4–3
Ion Exchange Process Detail

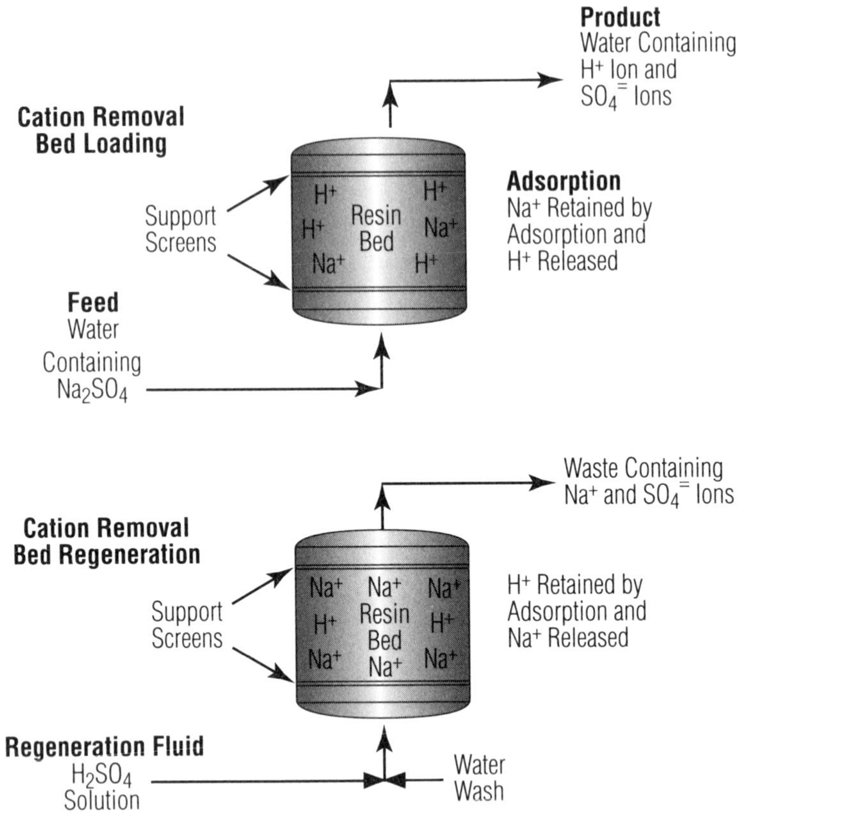

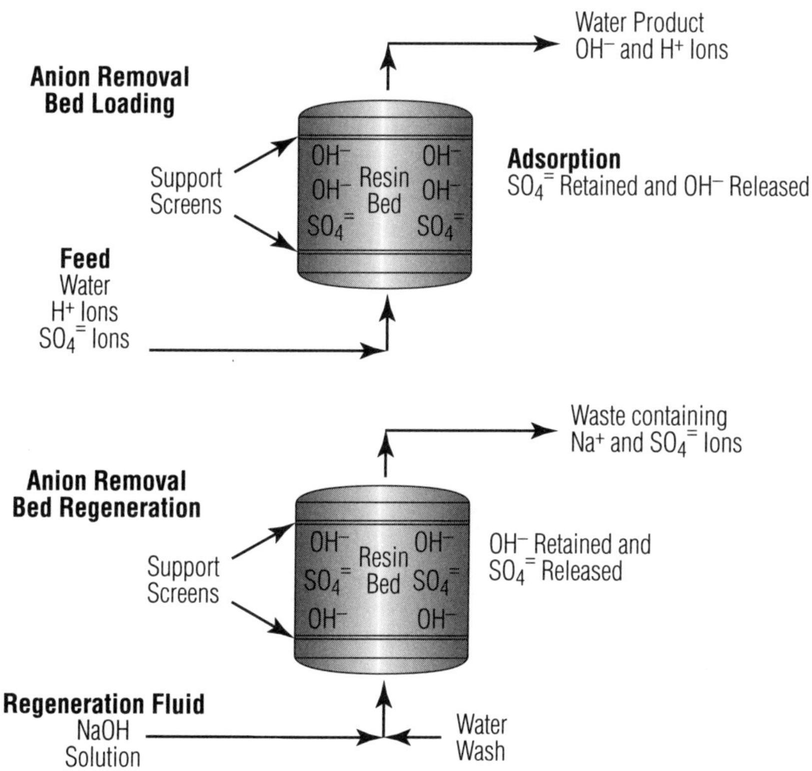

Figure 8–4–4
Ion Exchange Process Detail

Supporting Information

Figure 8–4–1 is a diagram of the five basic steps in the switch and regeneration of an ion-exchange bed. Figures 8–4–3 and 8–4–4 demonstrate the ion exchange that takes place in online service and in the regeneration cycles.

As can be seen, a number of possibilities exist for error in preparation of solutions, valving, and flow rates.

Figure 8–4–2 is a diagram of a set of ion-exchange beds. It requires a large number of automatic values and several key instruments to make up a fully automated system. There are a number of possible equipment problems that could be the cause of a variety of process upsets.

EXERCISE 8.5 HEAT EXCHANGER—TANK COMBINATION

A problem with a combination of equipment is very common. It can involve several areas and a number of process operators.

Good communications and teamwork by the operations team are required to document and analyze the effects that lead to probable cause of a process problem. The problem-solving capability of the operations team depends upon the information provided by the process operator to describe, analyze, and conclude the cause from effects.

As this exercise will demonstrate, the effects can be numerous and spread far throughout a number of unit operations. The probable cause is there, which connects them together and helps solve the problem.

1. A general layout is shown in Figure 8–5–4.
2. Process A, which is the responsibility of process operator A (Figure 8–5–3): A 500,000 gallon storage tank in Tank Farm A has discharged vapor through the emergency vent. This steam-heated, inert-blanketed tank contains a flammable material from Process B. The tank farm is process operator A's responsibility.
3. Process B, which is the responsibility of process operator B (Figures 8–5–1 and 8–5–2). Hydro C is produced at 120°C in a reactor. The product is cooled and sent to storage in Tank A. The reactor and cooler are operator B's responsibilities.
4. CCR operator (Figure 8–5–4): The CCR is informed by process operator A that Tank A is discharging through the emergency vent.

INSTRUCTIONS

1. Each team member decides upon the immediate action he should take and why. On the worksheet, describe all actions taken.
2. Each team studies the information and lists all the effects that could cause the emergency vent to fail.
3. Each team writes a definition of the problem.
4. The team lists possible causes and compares them to effects.
5. Each team selects the most probable cause and the recommended action.
6. Submit your write-up of the exercise to the instructor.

Table 8–5–1
Standard Operating Conditions

Cx flow (lb/hr)	F1	25,000
Hydrogen flow (lb/hr)	F2	2,000
Reactor exit temp. (°C)	T1	120
Hydro C cooler exit temp. (°C)	T2	30
Tank pressure minimum (psig PV1)		1.0
Tank pressure maximum (psig PV2)		4.0
Emergency vent setting		
	Maximum	5 psig
	Minimum	1 psig
Tank temperature (°C TT)		
	Maximum	40
	Minimum	15
Tank level (% LT)		
	Maximum	90
	Minimum	10
Tank pressure (psig)		
	Minimum PV1	1.0
	Maximum PV2	4.0
Emergency vent setting (psig)		
	Minimum	1 psig
	Maximum	5 psig
Hydro C properties		
Boiling point (°C)		80
Freezing point (°C)		10
Vapor press. (mm Hg)		
25°C		50
40°C		200
60°C		500
80°C		760
Specific gravity		0.80
Heat of vaporization (Btu/lb/°F)		300
Specific heat (Btu/lb/°F)		0.70

Table 8–5–2
Field Observations

Team A

T1	110°C
T2	45°C
PT stable	6 psig
TT	60°C
F1	23,000 pph
F2	1,800 pph
LT	50%

Team B

T1	130°C
T2	45°C
PT stable	6 psig
TT	45°C
F1	25,000 pph
F2	2,000 pph
LT	40%

Team C

T1	120°C
T2	35°C
PT stable	6 psig
TT	35°C
F1	20,000 pph
F2	16,000 pph
LT	60%

Supporting Information

A new unit is being brought online in another area. There have been a number of serious startup problems.

There is a caution tag on the tank access ladder which alerts all personnel to a possible release on an earlier shift. Operator A recalls seeing it mentioned in the shift log.

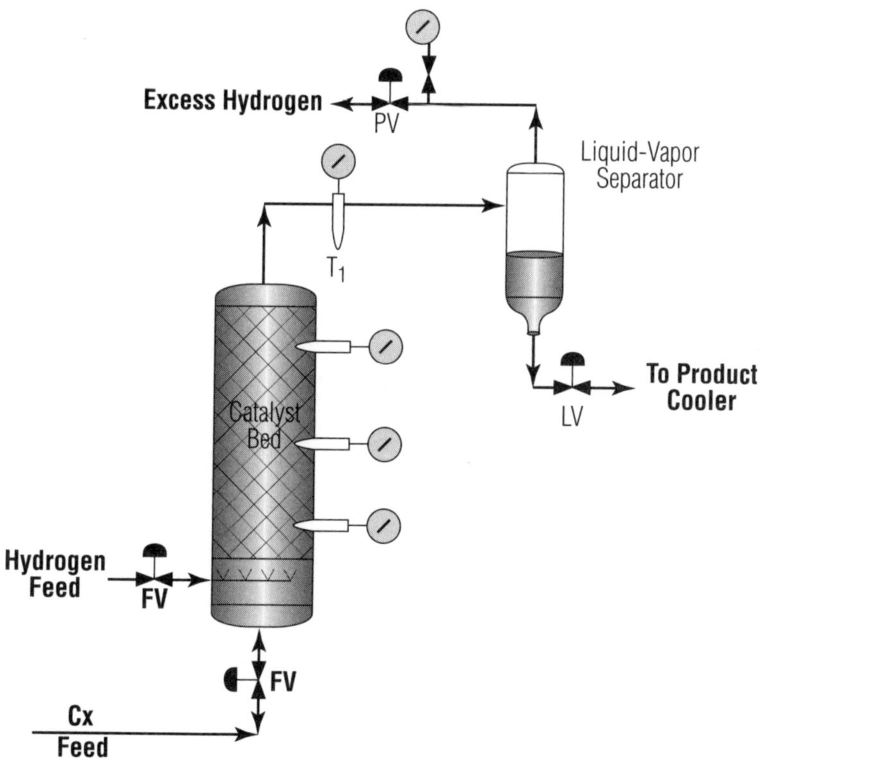

Figure 8–5–1
Hydro C Reactor

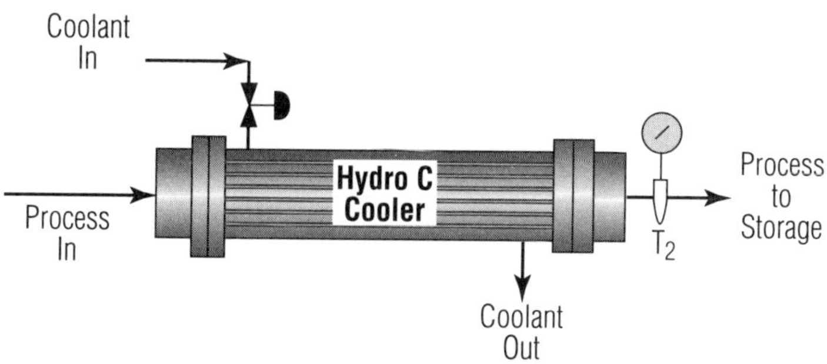

Figure 8–5–2
Hydro C Cooler

Figure 8–5–3
Typical Storage Tank and Accessories

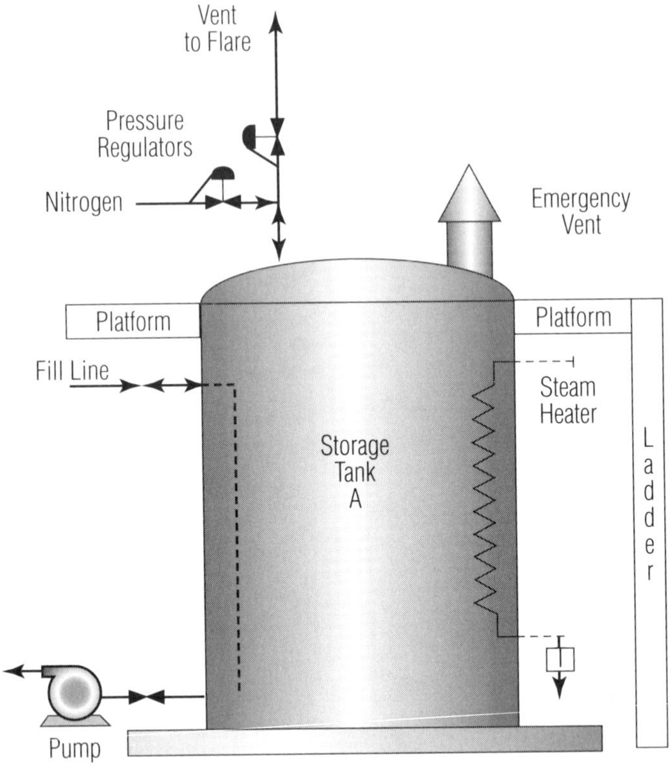

Figure 8–5–4
General Layout

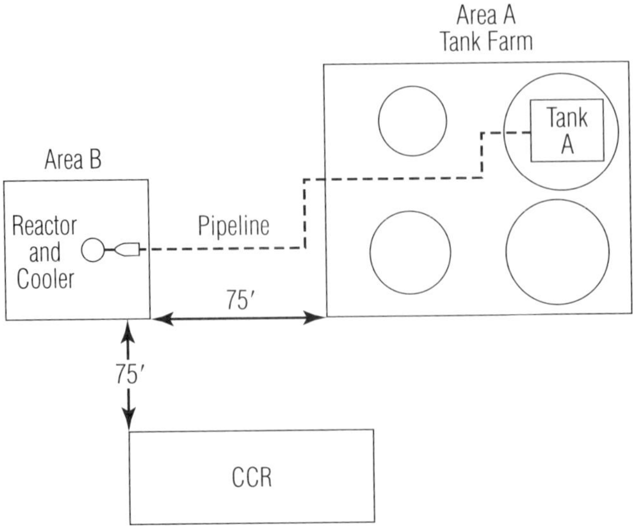

GLOSSARY

Action The final step in the problem-solving process. The change made to eliminate the root cause of a problem.

Analysis The deductive reasoning phase of the problem-solving process. The study of the effects of a problem with the objective to disclose all probable causes.

Cause The improper interaction among people, places, and things that results in a process problem, which is detected through abnormal effects.

Cause, Probable A potential but not yet proven root cause. Probable causes can be effects and effects can be probable causes until properly placed in a time frame of events.

Cause, Root The most probable cause that explains the effects. A root cause, if eliminated, eliminates the problem. May be referred to as the primary or direct cause.

Central Control Room A remote location with the major controls sufficient to run a chemical plant at steady state conditions.

Certification A product conforms to specifications tested by approved analytical procedures.

Checklist A list used as a reminder of tasks one is responsible for completing.

Communications Network An agreement on the exchange of information among those who need to know.

Conclusion The identity of the most probable or root cause from the list produced by analysis.

Contract Services Plant activities completed by outside workers under contract. The maintenance team is usually the coordinator.

Contract Worker Outside personnel used to complete selected work under a contract.

Corrective Action A change that is made to minimize the effects of a process upset and that is aimed at eliminating root cause.

Data Sheet A form designed for entry of specific process conditions.

Dead Time Wasted or unused time in an otherwise full schedule of planned activity.

Deadline A point in time when an objective is to be met.

Detection The initial discovery and communication on the effects of a problem.

Discovery The first step in the problem-solving process. Collect the effects of a problem into a logical order.

Dummy Run Mentally walking through a plant operation by simulation of the actual operation using non-toxic materials.

Effect A measured, observed, or sensed abnormality in the interaction of people, places, and/or things.

Event The result of an effect with a process upset being one extreme. *See* Incident.

Evolution The continuous change in petrochemical plant operations due to changes in technology and organization.

Fail-safe A pre-determined reaction of equipment in a safe direction to a process upset.

Feedback Response on the results of an action taken.

Feedstock The prime material fed to a process.

Follow-up Confirmation that an action taken produced the desired result.

Hazardous Materials Process materials with toxic, flammable, or other properties that are harmful to human health.

Housekeeping The activity needed to maintain a neat and orderly workplace.

Incident An event created by effects with a high potential for injury or equipment damage.

Installed Spare An exact duplicate of an equipment piece that permits a change in service if needed.

Interaction Two or more activities having an effect on each other that must be coordinated.

Intercom A speaker service between the central control room and the operators in process units.

Intrinsic Problem A cyclic activity that repeats in a continuous process, detracting from performance.

Log A written account of plant activity on a planned basis.

Management The act of organizing or controlling an activity at an acceptable level.

Mechanical The resources assigned to install, calibrate, and maintain facilities.

Network A set of circumstances or activities that are dependent upon one another for success.

Novice A beginner who must be trained on the job.

Oversight Overall responsibility to observe and report on a specific activity.

Performance Parameter A defined requirement in the output of an equipment piece.

Plant Patrol The hands-on comparison of what is to what should be in order to detect and prevent problems.

Plant Staff Overall management of plant resources to meet business objectives.

Prime Mover A term that refers to pumps and compressors used to generate fluid flow.

Probable Cause The most likely effect that initiated the problem.

Problem What is compared to what should be in the problem-solving process. Shows undesirable effects.

Problem, Acute An unpredicted change that has an immediate, negative effect on plant operation.

Problem, Chronic A predictable change in operations that penalizes productivity.

Problem, Expected The anticipated change the troubleshooter must monitor to prevent problems. *See* Intrinsic Problem

Problem-Solving Process A methodical exercise based on deductive reasoning that is applied to effects of a problem to expose the cause.

Process Condition A process variable that is controlled to achieve steady-state operation that produces predicted chemical and physical change.

Process Control Engineer Foremost technical assistance in day-to-day operations.

Process Control Laboratory Analytical service to operating units.

Process Operator The person responsible for continuous plant operations. May be called process technician.

Process Technology A combination of chemical and physical principles interacted to establish standard operating conditions necessary to complete physical and chemical changes.

Process Troubleshooting The detection and action taken to prevent and manage process problems

Process Upset A process unit operating unexpectedly outside standard operating conditions.

Process Variable A process condition that is controlled between predetermined limits to control process performance.

Procrastination Delay of a necessary activity until a future time.

Production The resources assigned to run the facilities and produce product.

Productivity A comparison of process performance to expectations, including utility and production rate.

Progressive Effects One effect becomes cause and creates a second effect, etc.

Reasoning, Deductive The use of effects to explain causes. The prediction of conditions to control results.

Reasoning, Inductive The use of past experience and training to explain the cause from the effects of a problem.

Relief Team Process operators scheduled to provide relief and work the next shift.

Resource Manpower, materials, and facilities available to carry out a specific task or mission.

Result The change produced by a deliberate action.

Root Cause The most probable effect behind a problem that, when eliminated, eliminates the problem.

Scientific Method Deduce conditions that should produce desired results.

Shift A rotated work period needed to cover continuous operations.

Shift Log A continuous record of plant operations.

Shift Team An organized group of process operators needed to run a multi-unit petrochemical process.

Short Cut The act of bypassing established procedures with the intent to reach an end result quickly.

Sketch A drawing without scale used to show a particular part of a process.

Solution The action taked to eliminate the cause of a problem.

Specification A condition or property of a product that can be measured to ensure it meets preset limits.

Standard Operating Condition (SOC) A process variable set to produce the desired process output.

Standard Operating Procedures (SOP) The steps necessary to operate equipment in order to maintain standard operating conditions.

Systemic Pertaining to a condition that consistently appears in a given system.

Team A group with common goals and a viable plan to meet them.

Team, Mechanical Responsible for the maintenance of plant facilities, including equipment, instruments, and electrical.

Team, Operations The unified resources of the mechanical, production, and technical teams.

Team Player One who works well with others to achieve teamwork.

Team, Production Responsible for the operation of plant facilities to meet business objectives.

Team, Technical The expertise needed to support the technology needs of the production and mechanical teams.

Team, Troubleshooting A group of people selected for their knowledge and skills who observe, document, analyze, and correct abnormal process conditions.

Teamwork A shared effort by knowledgeable and skillful people to reach a common goal.

Technical The resources assigned to support the technology needs of the production and mechanical teams.

Technical Specialist An expert in one or more aspects of process technology.

Timetable A checklist organized against time. Activities set down from first to last.

Trial and Error Repeated change in a condition until the desired result appears.

Troubleshoot An organized effort to prevent process upsets through discovery of abnormal effects, analysis of abnormal effects for cause, and taking of action to eliminate cause.

Troubleshooter One who seeks out unwanted process change and the causes behind it.

Troubleshooting Team A group of people selected for their knowledge and skills who observe, document, analyze, and correct abnormal process conditions.

Unexpected Problem A change where normal work practices did not detect the effects.

Utility The hours available to produce product compared to total hours.

Verbal Communications Communication by word of mouth.

Workmanship The quality of both mental and physical activity in a professional manner.

Work Practice An accepted way or method to complete a defined task.

Workstation A computer designed to carry out a specific agenda of assistance to operations.

Yield The useful recovery of product from feedstock.

INDEX

S

T

U

V

W